THE GRAPHENE
REVOLUTION

THE GRAPHENE REVOLUTION

The Weird Science of the Ultrathin

BRIAN CLEGG

ICON

Published in the UK in 2018
by Icon Books Ltd, Omnibus Business Centre,
39–41 North Road, London N7 9DP
email: info@iconbooks.com
www.iconbooks.com

Sold in the UK, Europe and Asia
by Faber & Faber Ltd, Bloomsbury House,
74–77 Great Russell Street,
London WC1B 3DA or their agents

Distributed in the UK, Europe and Asia
by Grantham Book Services,
Trent Road, Grantham NG31 7XQ

Distributed in the USA
by Publishers Group West,
1700 Fourth Street, Berkeley, CA 94710

Distributed in Australia and New Zealand
by Allen & Unwin Pty Ltd,
PO Box 8500, 83 Alexander Street,
Crows Nest, NSW 2065

Distributed in South Africa
by Jonathan Ball, Office B4, The District,
41 Sir Lowry Road, Woodstock 7925

Distributed in India by Penguin Books India,
7th Floor, Infinity Tower – C, DLF Cyber City,
Gurgaon 122002, Haryana

Distributed in Canada by Publishers Group Canada,
76 Stafford Street, Unit 300
Toronto, Ontario M6J 2S1

ISBN: 978-178578-376-0

Typeset in Iowan by Marie Doherty

Printed and bound in Great Britain by
Clays Ltd, Elcograf S.p.A.

ABOUT THE AUTHOR

Brian Clegg's most recent books are *The Reality Frame* (Icon, 2017) and *Are Numbers Real?* (St Martin's Press, 2017); he has also written *Gravitational Waves* and *Big Data* in the Hot Science series. His *Dice World* and *A Brief History of Infinity* were both longlisted for the Royal Society Prize for Science Books. Brian has written for numerous publications including *The Wall Street Journal*, *Nature*, *BBC Focus*, *Physics World*, *The Times*, *The Observer*, *Good Housekeeping* and *Playboy*. He is the editor of popularscience.co.uk and blogs at brianclegg .blogspot.com.

www.brianclegg.net

For Gillian, Rebecca and Chelsea

ACKNOWLEDGEMENTS

My thanks to the team at Icon Books involved in producing this series, notably Duncan Heath, Simon Flynn, Robert Sharman and Andrew Furlow. Although their names will crop up a lot in this book, it's also not possible to talk about ultrathin materials without thanking physicists Andre Geim and Konstantin Novoselov for starting this whole business.

CONTENTS

THE STICKY TAPE SOLUTION 1

Big science–little science

There was a time when a lone scientist, or a handful of individuals working in a lab on a shoestring budget, could achieve wonderful things. Think of pretty well any scientific discovery that was made before the Second World War and you'll find that neither finance nor staffing were huge. However, it would be easy to think that the hot science subjects of the 21st century all require massive budgets and enormous teams. This book is about a subject that has shattered this assumption.

Think back to the major announcements that were made in science since 2000. Early on in the century, the Human Genome Project published its results, with drafts from both the $3 billion public programme and the $300 million private Celera programme released jointly in 2001 after many years of work. In 2013, the team working on the Large Hadron Collider at CERN near Geneva, often called 'the biggest machine in the world', announced the discovery of a particle consistent with a Higgs boson. The collider and the staff working on it have cost over $5 billion to date.

Similarly, in 2016 and 2017 we have had a number of announcements of discoveries of gravitational waves made from the LIGO observatories, built at a cost of over $1 billion and with over 1,000 scientists worldwide involved in the project. And all of this is dwarfed by the funding that has been piled into the International Space Station which has cost over $100 billion without a single major scientific discovery to its name.*

So, what could two physicists working in Manchester, England achieve with a negligible budget, some blocks of graphite and a few rolls of sticky tape? It would turn out to be rather a lot – perhaps the most far-reaching technological breakthrough of the 21st century to date. The development of ultrathin materials that emerged from the Manchester work has far greater practical value than any of the multi-billion dollar experiments named above, yet also contributes major steps forward in our understanding of both physics and chemistry. This is big-impact small science on a minimal budget.

The city of Manchester has a strong reputation for scientific discovery – particularly when working on the atomic scale. It was there that John Dalton put forward his atomic

---

* To be fair, the primary role of the International Space Station is not scientific experimentation but to provide a test bed for approaches to space exploration, so arguably it shouldn't be considered a science project. It is, however, often misleadingly portrayed as a scientific endeavour.

theory that transformed our understanding of matter in the early 1800s. Nearly a century on, in 1900, Owens College in Manchester, soon to become part of the Victoria University of Manchester, saw the opening of an all-new physics building, a state-of-the-art facility, complete with a remarkably modern ventilation system which used oil baths to remove the soot from the smoky atmosphere of the country's leading industrial city.

It was in this laboratory that Ernest Rutherford discovered the structure of the atom and Niels Bohr made the first steps towards a quantum mechanical understanding of atomic structure. Since then, all manner of scientific developments have followed in Manchester, from the construction of the Jodrell Bank radio astronomy observatory to Alan Turing's work on computing. And it was that same Manchester University physics department* that unwittingly played host to Andre Geim and Konstantin (Kostya) Novoselov's 'Friday night experiments' which led to the discovery of the new wonder material, graphene, followed by work on a range of other ultrathin substances.

These two Russian-born physicists first met when Novoselov was supervised by Geim on his PhD – which he was awarded in 2004 – at the Radboud University of Nijmegen in the Netherlands. Sixteen years older, Geim had by then already gained a considerable reputation for original science combined with quirkiness and lateral thinking. Nothing shows this more clearly than his use of both frogs and a hamster in his work.

---

* Although not the same building – the 1900 building is now an administration block.

Levitating frogs and co-authoring with a hamster – the quirky history of graphene's discoverers

In 2000, ten years before he won the Nobel Prize for his work on graphene, Andre Geim won the Ig Nobel Prize for levitating frogs.* The Ig Nobel is a humorous award that has been presented since 1991 for scientific research that 'first makes people laugh, then makes them think'.** It is an entertaining reflection on strange-sounding research and has been won by scientific papers and inventions with citations such as 'Can a cat be both a solid and a liquid?', 'Dung Beetles use the Milky Way for Orientation' and 'Determining the ideal density of airborne wasabi (pungent horseradish) to awaken sleeping people in case of fire or other emergency.' What Geim demonstrated was that magnetic levitation of living organisms – and particularly frogs – was perfectly possible.

Anyone who has played with a pair of magnets knows that when they are aligned north pole to north pole, or south pole to south pole, they repel each other. If strong enough magnets are kept in alignment, one can be made to hover above the other. This clearly has potential practical applications. The idea that the repulsion effect could be used to get a train to hover over its tracks has been around since the start of the 20th century and a number of prototype 'maglev' (magnetic levitation) trains have run over the years. However, large-scale commercial application is only just

---

* Geim is the first person ever to win both an Ig Nobel and a Nobel prize.
** The Ig Nobel was started by Marc Abrahams – see www.improbable .com for more details.

becoming feasible with the development of ultra-powerful superconducting magnets. The Chuo Shinkansen line in Japan, which is expected to run at speeds of up to 500 miles per hour, is under construction at the time of writing.

To get many tonnes of train to float above the rails requires a lot of power – but it involves conventional magnetic repulsion between a magnet and pieces of metal, in which the magnet induces a magnetic field, a process familiar to Michael Faraday. But we all know that magnets don't work on living things – so how could Geim, with his then collaborator Michael Berry of the University of Bristol, get a frog to float in mid-air using only a powerful electromagnet?

It's worth thinking first about the way that different metals react to magnets. Iron, for example, has a strong response to magnetism, while copper – which like iron is a good electrical conductor – does not. This primarily reflects the way the electrons are grouped around the atoms of these metals. As we will see in more detail on page 33, atoms have 'shells' occupied by electrons, and each shell has a limited capacity. Copper has a single electron in its outer shell, which can be easily detached to conduct electricity, but this leaves a full outer shell, which means that the copper atoms in the lattice structure that make up a piece of the metal are relatively symmetrical. When iron loses an electron for conduction, though, its outer shell is not full – this means there's a degree of asymmetry in the atoms, and each atom can act like a tiny magnet, lining up under the influence of a magnetic field.

Frogs and other living things, by contrast, aren't made of metals (apart from small amounts in the blood etc.)

– but frogs *are* made up of atoms and molecules which have attached electrons. Particularly handy is the asymmetric structure of water. In a strong magnetic field, water molecules will tend to line up and oppose the initial field that is acting on them, forming a weak magnet, a phenomenon known as diamagnetism. The effect is billions of times smaller than the field induced in a magnetic metal, but if the magnet influencing the frog is strong enough, the effect is sufficient to overcome the remarkably weak force of gravity.*

A spare-time look at an unlikely interaction between magnets and water was the original stimulus for Geim to begin his work on frogs. He notes that it had been claimed for some time that putting magnets on taps and water pipes would prevent a build-up of limescale (and indeed many products are available online which claim to do just this). But it was hard to understand why they would work and many suspected that they were just ways to make easy money. As Geim put it: 'The physics behind [the action] remains unclear, and many researchers are sceptical about the very existence of the effect.' Never one to be put off by opinion, Geim attempted several unsuccessful experiments on the effect and eventually commented that he still had nothing to add to the argument. But the process got him thinking laterally about water, particularly as his day job involved working with extremely powerful magnets – that is, magnets

---

* Gravity may seem powerful, but compare the gravitational influence of the whole Earth trying to pull a fridge magnet off the fridge and the tiny bit of magnet that is holding it in place. The magnet wins. Gravity is around a trillion, trillion, trillion times weaker than electromagnetism.

200 times stronger than a typical modern high-strength neodymium magnet.

Frogs were chosen as the subjects for the experiment because they are light, have a high water content as animals go, and because, to put it bluntly, there would have been less of a problem if they had ended up exploding than there would from experimenting on people or puppies. And it was much easier to build a magnetic 'cup' using electromagnets to support a frog in mid-air than it would have been to levitate a human. The first subjects for the experiment, though, were less likely to catch the attention of the media – simple drops of water. Geim noted in his Nobel lecture: 'Pouring water into one's equipment is certainly not a standard approach, and I cannot recall why I behaved so "unprofessionally". Apparently no one had tried such a silly thing before ...' His colleagues immediately suggested he could try the experiment out on drops of beer as a follow-up experiment. It is not recorded whether Geim gave this a try, though it would have been entirely in character for him to do so.

A more realistic concern once moving on from globules of liquid to living things is that the extremely strong magnetic field would set up electrical currents in the brain. This is observed in a medical process known as transcranial magnetic stimulation. Powerful magnets positioned near to the skull do have a significant effect, starting electrical currents flowing in the cranial tissue. At low levels these can have beneficial effects, and are useful to apply a non-invasive prod to the brain, but with a strong enough field, the induced currents can cause seizures. However, the frogs seemed unharmed in the experiment.

Although Berry and Geim's study was a serious piece of work (and deals with diamagnetic* materials in general, not just the headline floating frogs), some of Geim's sense of humour still managed to creep into the mostly very sober paper that the pair wrote on the subject. Here we are told: '[Using the effect on living organisms] could cause strange sensations; for example, if [the magnetic susceptibility** of flesh is greater than the magnetic susceptibility of] bone, the creature would be suspended by its flesh with its bones hanging down inside, in a bizarre reversal of the usual situation that could inspire a new (and expensive) type of face-lift.'

The following year, Geim confirmed his reputation for injecting levity into what was otherwise serious work when he wrote a paper for the very straitlaced *Physica B: Condensed Matter* journal on 'Detection of earth rotation with a diamagnetically levitating gyroscope' (a more practical application of levitation). His single co-author for this paper was named as H.A.M.S. ter Tisha – in other words, his pet hamster, Tisha.***

In a way, this recognition was an opportunity to reward Tisha for the animal's contribution to earlier levitation research efforts. Tisha had been the first test subject for

---

* Diamagnetic materials are substances that are repelled by magnetic fields despite not being themselves magnetic materials.

** Magnetic susceptibility is a measure of how much a substance is attracted or repelled by a magnetic field.

*** Tisha was not the only animal in history to co-author a scientific paper. For example, American biologist Polly Matzinger co-authored with her dog Galadriel Mirkwood, while the American physicist Jack Hetherington not only co-authored with his cat, Chester (operating under the name F.D.C. Willard), Chester even has a byline for writing a solo article for a French popular science magazine.

living organism levitation, but had appeared distressed by the experience. As Geim would later put it: 'First we used a hamster. After we saw the hamster didn't like it, we took a frog.' The frog, it seems, was more phlegmatic about the whole thing.

Pencil thin

The work that won Geim and Novoselov their Nobel Prize was on graphene, a sheet of graphite – one of the crystalline forms of carbon – just one atom thick. Graphite is a substance that we have all used at some time, though an oddity of history means that we are more likely to call it 'lead' when it appears down the centre of a pencil.

At first glance, it isn't obvious why anyone should associate graphite with lead. Lead is a dull grey metal, obviously not a material for drawing with, while graphite is carbon in the form of a black, shiny non-metallic material, not unlike coal in appearance. Sometimes, rather dubious guesswork has been used to come up with an explanation of why we call the writing bit of a pencil its lead. Perhaps the most plausible (if incorrect) suggestion is that it was because Romans wrote with a stylus that was made out of lead.

The better substantiated answer makes rather more sense. The natural lead ore galena is lead sulfide. (Because of impurities, as well as being the main source of lead, galena is also where much of our silver comes from.) Galena is a shiny black crystalline substance, which has a strong resemblance to naturally occurring graphite crystals.

When graphite was first discovered it was called plumbago or black lead, because it was actually thought to be a variant of galena. Though the distinction was pretty much cleared up by the 1770s, the name 'lead' stuck as the way we refer to pencil cores.

It should be fairly obvious that graphite is a useful material to make pencils from, even if you've never seen any. Its very name, coined in 1789 by German geologist Abraham Werner (as 'graphit', without an e on the end), labels it a 'writing mineral'. England had a near-monopoly on good quality pencils (which were originally made by wrapping a stick sawn from a block of graphite in string or animal skin to strengthen it), as the only known large-scale deposit of high quality graphite in Europe was in Cumbria, in the north of the country.* It was only when other countries began to use the more easily obtained powdered graphite that pencils became common worldwide.

The essence of graphite's effectiveness as the writing material in a pencil is its crystalline structure. We're more likely to think of diamond as a crystalline form of carbon (which it is), as we're used to crystals being transparent and hard. But any element or compound with a regular repeating structure, where the atoms are bound together in a lattice, is a crystal. Metals, for example, despite being very different in looks to diamonds, are also crystals. And the graphite in a pencil lead is just as much a crystal as is diamond – but based on an alternative arrangement of the atoms.

---

* Hence the location of the UK's foremost pencil museum at Keswick in Cumbria.

Rather than being a homogenous solid like a diamond, a graphite crystal is built up of layer upon layer of atom-thin sheets. These sheets are very strong in the plane of the sheet, but are only lightly attached to each other, and so easily slip over each other when put under pressure. The action of writing with a pencil rubs the graphite tip on to a sheet of paper, forcing sheets of graphite to slide off the tip of the pencil and be deposited on the paper. This ease of movement of the layers over each other also accounts for the oddity that this solid material makes a good lubricant. You'll often find lubricants that involve graphite because of the ease with which the layers slip over each other.

It was exactly this kind of mechanism – the ease with which layers can be removed from a block of graphite – that would be used to produce graphene. Because graphene is nothing more than a single atomic layer from a block of graphite. But it was necessary to go considerably further than simply rubbing a pencil on some paper. The layer of graphite this leaves behind on the paper consists of many graphene sheets – this has to be the case, or you wouldn't be able to see what you'd written; graphene itself is transparent. How Geim and Novoselov were able to produce this remarkable substance owes much to a happy coincidence.

Playing in Manchester

In 2001, Andre Geim moved from his previous post in the Netherlands to Manchester, to take up a position as professor of physics. As he had demonstrated with the levitating frog,

Geim likes to treat science as an adventure that can head off in any possible direction. He claims that one of the obstacles to the kind of work he likes to do is 'the typical academic'. Such creatures he defines as 'people who are put on the rails like a train by their supervisor and they continue doing all the same stuff from their scientific cradles to their scientific coffins. They all go along the same straight rail line – not a British rail line, but a straight rail line like in Siberia. I know plenty of Russians and British who do exactly the same thing without trying to move sideways because it's danger-ous, because it's not what our instincts tell us ... When you move from place to place you learn different things and this [gives you] pieces for your Lego game. The more pieces you have, the more complex the structures you can make.'

This reference to Lego building bricks reflects an import-ant part of the approach that has come to typify Geim's work. His 'Lego doctrine' is to see what you've got available in a lab – the Lego pieces – and try to do something new and dif-ferent with it, assembling the pieces to make new models. In his railway line analogy, this involves veering off the straight line into the green-field sites around it. His view of the Lego approach is that 'You have all these different pieces and you have to build something based strictly on the pieces you've got.' One of Geim's ways to counter the limitations of the traditional railway line was to encourage free thinking time on a Friday night, supporting his belief that there should be 'search not re-search'.

Geim's unusual approach dates back to his experience as a PhD student at the Institute of Solid State Physics in Chernogolovka, near Moscow. His thesis there was on

'Investigations of mechanisms of transport relaxation in metals by a helicon resistance method', which he ruefully admitted in his Nobel Prize lecture 'was as interesting at that time as it sounds to the reader today'. Geim noted that his total of five journal papers on the subject, plus his thesis, have only been cited twice – and then only by co-authors. 'The subject was dead a decade before I even started my PhD. However, every cloud has its silver lining, and what I uniquely learned from that experience was that I should never torture research students by offering them "zombie" projects.'

Bringing to life a zombie project was not Geim's intention when he asked a Manchester PhD student, Da Jiang, to take a chunk of graphite and to make from it the thinnest sheet that he could achieve. Despite newspaper articles to the contrary, the graphite used was not a piece of pencil lead, but a type known as highly oriented pyrolytic – a carefully produced block of near-perfect carbon layers produced by heating carbon sources in high temperatures and pressures,* costing around £300 a block. At least, the intention was to give Da a piece of highly oriented pyrolytic graphite. Geim admitted in his Nobel lecture that he unintentionally gave Da a block of high-density graphite instead, which makes it far harder to produce thin, uniform layers. Geim hoped that Da could produce a sample so thin that it would act more like a two-dimensional atomic lattice, which was predicted to have quite different properties to ordinary graphite.

---

* One good source of high-quality graphite for producing graphene is so-called Kish graphite, which is deposited on the surface of molten iron during the steel-making process.

Da approached this task by grinding away layers until he had produced the thinnest slice of graphite that the equipment he was using could leave behind – but the sliver of carbon was not near enough to the atom-scale thinness that Geim knew was required to achieve the special properties he hoped to explore. This was a typical example of Geim pushing the boundaries: many theorists at the time thought that a two-dimensional crystal, such as the theoretical atom-thin sheet of graphite that was graphene, would be inherently unstable if separated from a block and would disintegrate to dust.

It seems that after failing, Da asked for a second piece of graphite to have another go – something that went beyond the practically non-existent budget of what Geim had by then christened his Friday night experiments, spare-time ventures into new directions, inspired by the success of his levitation work. As Geim drily put it, when Da asked for another block of graphite, 'You can imagine how excited I was.' But rather than send more good money after bad, Geim followed up an idea suggested by his post-doctoral student, Oleg Shklyarevskii.

It seems that Shklyarevskii overheard Geim moaning about Da's disappearing graphite block, which Geim compared to polishing an expensive mountain to get a grain of sand. Shklyarevskii pointed out that scientists who worked with blocks of graphite needed to clean them first, to make sure the surfaces were even. Shklyarevskii was an expert in scanning tunnelling microscopy and in that field, blocks of graphite were (and are) used as a standard reference sample to set up the microscope.

To make a pristine sample for the microscope's calibration, the scientists applied strips of Scotch tape to the surface of the block and peeled them back off, taking away a thin layer of the carbon and leaving behind a smooth, clean surface. The pieces of tape were then binned. Geim observed: 'What these guys did not realise was that throwing away the Scotch tape they were throwing away the Nobel Prize as well.'

In true Friday night experiment style, Geim and Shklyarevskii looked through their colleagues' waste baskets and found plenty of pieces of tape which had graphite layers left behind on them. (Thankfully, solid state physics lab bins tend to be less hazardous than those in biology or chemistry labs.) Though the layers of graphite on the tape strips still weren't thin enough to have the potentially remarkable properties of graphene, under the microscope, the samples proved to be thinner than anything Da had achieved using the grinding method. Some of the deposits of graphite on the tape were so thin that they were transparent – which meant that there could only be a small number of graphene layers present.

'We did not invent graphene,' Geim says, 'we only saw what was laid up for five hundred years under our noses.'*

By this time, Konstantin Novoselov was fully committed to the project with Geim, and together the pair worked on

---

* The name 'graphene' was not dreamed up by Geim, but had been in use since the 1980s to describe the layers in a piece of graphite and that also formed tiny carbon 'nanotubes' when in a more stable rolled structure. But no one believed that a flat sheet of graphene could be separated or that it would remain as a stable substance.

the separation of thin sheets of graphite, repeatedly peeling them off with adhesive tapes, then pressing the tapes on to an oxidised silicon wafer. The closest layer of graphene tended to stick to the silicon oxide as a result of a weak inter-atomic force called a van der Waals force (of which much more later). When the tape was carefully peeled away from the silicon wafer, a much thinner layer would be left behind. After a year of experimenting with different approaches, the pair managed to reach their ultimate goal – a layer of carbon a single atom deep, forming a neat lattice of atoms like an interlocked set of hexagonal benzene rings but with only carbon present, which would prove to be remarkably versatile.

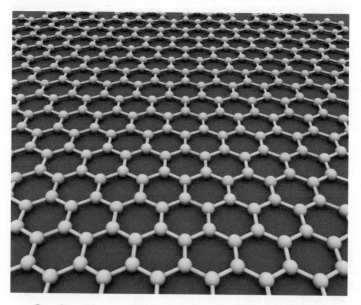

Graphene forms a lattice of hexagonal carbon rings.

The mere fact that this near-two-dimensional sheet existed at all hinted at its unusual nature. All atoms are constantly on the move, and in normal circumstances it is usually the larger structure of a body of matter that holds a thin layer together. Graphite is fine, as each layer supports the next, but you might expect a single layer of atoms to pull itself apart from the thermal activity of the atoms alone. However, this new material was strong enough to resist those rippling forces. Graphene is both stronger than steel by a factor of 100, and 100 times more conductive than copper.

Geim and Novoselov's work has enabled them in just a handful of years since their discovery to open up possibilities from super-strong materials and flexible atom-thin electronics to molecular sieves to purify seawater. But before we can understand graphene and its amazing promise, we need to start with the ultimate Lego bricks – the atoms that make it up – and how those atoms interact.

THE ESSENCE OF MATTER 2

Atoms everywhere

To discover where the concept of atoms came from, we need to go back around 2,500 years in time. The Ancient Greeks had two main competing theories on the nature of matter. In the 5th century BC, Empedocles, a philosopher based in the Sicilian city of Agrigentum, introduced the concept of the four elements: earth, water, air and fire. Nearly 100 years later, Aristotle, probably the most famous of all the Greek philosophers, added a fifth element called quintessence (also sometimes called the ether). This, he felt, was necessary as the dominant cosmology of the time considered everything above the orbit of the Moon to be unchanging and eternal – requiring something more stable than the Earthly elements.

The competitor theory dated back to around the same time that Empedocles was coming up with the elements. Another philosopher, Democritus, supported by his teacher Leucippus (if he existed*), suggested that everything was made up of atoms – literally 'uncuttable'** fragments of matter, the smallest possible components of stuff.

In principle, the two theories were compatible – just as now we are quite happy to talk about both the elements, which reflect the different structures of atom, and of atoms themselves. Philosophically, though – which is the only thing that mattered in Ancient Greece – the two theories had significant differences of approach. Atomists thought that each different type of substance had its own specific variant of atom that made it the material that it was. So, wood atoms would be different from cheese atoms – each was assumed to have a distinctive shape. Those who supported the theory of the four/five elements not only rejected this, but from Aristotle onwards were unhappy with the whole concept of atoms because of something they implied.

The problem with atoms, as this theory imagined them, was that if they did exist there must also be empty space in between them. Relatively few three-dimensional shapes can fill all of space without leaving gaps of nothingness in between. There certainly aren't enough such shapes for there to be one for each type of matter. But Aristotle was

---

* There is very little evidence for the life of Leucippus, who preceded Democritus, and some have suggested he was a fictional means to add weight to the theory of atoms.

** The original word is a combination of *a* meaning 'not' and *tomos* for (roughly) a cut or cutting. It came to us via the Latin *atomus*.

convinced that nature did not support the existence of a void or vacuum.

One of Aristotle's arguments against the existence of the void was remarkably like Newton's first law of motion. Aristotle commented that if a void existed, 'No one could say why something moved will come to rest somewhere; why should it do so here rather than there? Hence it will either remain at rest or must move on to infinity unless something stronger hinders it.' And he was quite right – it does. He just thought that this was nonsense, because in the normal world, such behaviour is usually restricted by friction and air resistance.

After a degree of toing and froing, it was the atomic theory that lost out among the Ancient Greek philosophers. This is not as shocking as it sounds. Perhaps surprisingly to us now, the five elements were a more useful scientific theory than was the version of atomic theory proposed by Democritus and Leucippus. We now know that Aristotle was wrong – but at least his theory made predictions about the way different materials behaved.

For example, if you burn a piece of wood it gives off the air-like smoke plus fiery flames and, if it's green wood, it will exude watery liquids as it is heated. You are left with earth-like ash. The element theory seemed to explain this, assuming wood was made up of a combination of the elements. By comparison, the Ancient Greek atomic theory didn't stand up to experiment. By allowing every type of material to have its own kind of atom, it didn't really reflect how one substance could be transformed into another. Consequently, Aristotle's five elements were to hold sway until Newton's time and beyond.

Dalton's discovery

By 1800, though, new discoveries were starting add to in far more elements than the original four or five, and the way that those elements combined in specific ways seemed to imply that the elements were themselves constructed from fundamental building blocks. John Dalton, who we've already met as a leading light of the early Manchester science scene, suggested that some matter was made up of multiple atoms of one kind of element, while other, more complex, substances combined different types of atoms in what we'd now call compounds.

Dalton was a remarkable character. He was a Quaker, which made it impossible for him to attend an English university, with places then only available to members of the Church of England. What education he had came primarily from reading and from the instruction of those around him. He appears to have been quite confident of his knowledge, as he was already acting as a schoolteacher by the time he was fifteen. It's not at all clear how he came by his atomic theory, though there is some evidence that it was influenced by his studies of different gases and liquids and the way that they interact.

Where the Ancient Greeks had been prepared to dream up different shapes for the atoms, Dalton had no concept of what an atom was like. Even though his theory was widely adopted, because it proved immensely useful, many of his contemporaries did not even believe that atoms per se really existed, but rather thought that they provided a useful model to describe how the different elements interacted. It wasn't until the early 20th century that there would be widespread acceptance that atoms truly existed.

At the heart of Dalton's big idea was the concept of atomic weight. He gave relative weights to different elements, starting with the lightest – hydrogen – which was allocated a weight of 1. Each element's atom was a distinct building block of matter that had its own weight (a multiple of that of hydrogen) and an individual tendency to react with other atoms to form compounds in simple ratios, reflected in whole numbers of different atoms.

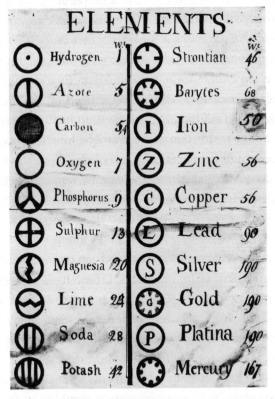

Elements and their atomic weights, from Dalton's
A New System of Chemical Philosophy.

To the modern mind, familiar with subatomic particles, it might seem obvious that there was something significant about the way Dalton's atomic weights were exactly multiples of the weight of hydrogen. Mathematically, when a series of items are all multiplies of the same value, there is an implication of an underlying numerical structure – and in the case of atoms, it's easy for us to think that Dalton must surely have seen that this implied either that all the atoms were made of hydrogen, or that all atoms had some other set of internal components – but this doesn't seem to have occurred to Dalton or his contemporaries.

Manchester historian of science James Sumner points out that Dalton did not actually believe that his atomic weights were exact multiples. When describing water as combining hydrogen and oxygen, Dalton notes that these atoms' weights 'are as 1:7, nearly ...'. Furthermore, says Sumner, 'This use of "nearly" recurs in many of Dalton's proposed proportions, and would perhaps have diverted attention away from any thoughts of a compositional principle.'

What's more, 'Dalton was strongly concerned with the physical *sizes* of the atoms, assuming them to be something like spheres: conceptualising an oxygen atom as somehow like a grouping of seven(ish) hydrogen atoms would be of no explanatory help here.' In practice, it seems that Dalton's concern was not so much to explore the fundamental structure of matter as to produce a pragmatic mechanism, based on the idea of component particles, to explain the behaviour of gases when mixed together. From this he was able to speculate on how elements combine. The atomic theory emerged from what Dalton had only originally intended to have a significantly smaller application.

Dalton's precise theory suffered from plenty of errors, based as it was on relatively crude measurements. His atomic weights were often inaccurate and he didn't realise that, for example, oxygen in the air was in the form of a molecule with a pair of atoms held together by chemical bonds. His ideas for the number of atoms in molecules were often well away from the values we now expect. There was no good way to determine what the ratio of the elements was and Dalton worked on a self-imposed 'rule of greatest simplicity' – which had no evidence to back it up. He merely assumed that the simplest possible combination of elements was correct unless there was something to suggest otherwise.

Dalton thought, for example, that water, which the familiar modern chemical formula of H_2O identifies as two hydrogen atoms to one atom of oxygen, was a 'binary' atom with one atom each of hydrogen and oxygen. Similarly, he thought that ammonia (NH_3) had one atom of azote (nitrogen) to one of hydrogen, and, most dramatically, he was far from the mark in suggesting that alcohol, presumably the ethanol (C_2H_6O) of alcoholic drinks, consisted of three carbon atoms and one atom of nitrogen.

He wouldn't, incidentally, have been happy with representations such as H_2O. He disliked this approach, introduced by the Swedish chemist Jöns Jacob Berzelius, and always used his own symbols, which in a style reminiscent of the Ancient Greek atomic concept, portrayed each element as a circle with a different shape inside it.

We shouldn't be too hard on Dalton, though, for his inability to spot the true proportions of elements in compounds. Not only did he have very limited equipment, even by the standards of 1800, he was working at the leading edge of scientific discovery for the day. Many of the elements he was working with had only been identified as distinct substances during the preceding three decades. Oxygen and nitrogen, for example, were first identified in the 1770s. And it was not until Einstein wrote a paper in 1905 on Brownian motion – the way that small particles such as those ejected from pollen grains bounce around in water as they are buffeted by the water molecules – that there was good evidence to suggest that atoms and molecules were real things.

No one had been sure what caused Brownian motion, named after the Scottish botanist Robert Brown.* It had even been thought for a while that it might be due to the life force in the pollen grains, until it was pointed out that it occurs with totally inanimate matter. But Einstein suggested that the motion was due to the constantly moving water molecules repeatedly bashing into the much larger particles, collectively causing the tiny specks to undergo a random dance** through the liquid. What convinced many that this was more than mere speculation was that Einstein backed up the idea with mathematics, showing that the observed behaviour was just what would be expected if molecules (and hence atoms) really existed, as described by Dalton's theory.

---

* Brown got the glory, even though the Dutch biologist Jan Ingenhousz had observed something similar with charcoal grains in water a good 50 years earlier.
** Traditionally given the non-PC description of a 'drunkard's walk'.

Not uncuttable

By the start of the 20th century, just as atoms were coming to be accepted as real things, evidence was showing up that atoms really didn't live up to their name as being 'uncuttable'. It all started with the unlikely discovery of cathode rays. The term may raise vague memories of old TV sets and computer monitors – the ones with the big lump sticking out at the back. Found in most homes up to the 1990s, these made use of what was essentially the same technology as that used by the likes of Victorian British physicist William Crookes. Crookes was a self-taught scientist who did early work on electrical effects in vacuum tubes, sealed glass tubes with most of the air pumped out. These tubes were commonly referred to as Crookes tubes.

Experimenters had been putting electrical charge across two electrodes inside such a partially evacuated tube since Michael Faraday noticed this produced a strange glow in the 1830s, but as better vacuum pumps became available, it was possible to remove most of the gas from the tube and the result was that the bulk of the tube went dark, but something invisible passed down the tube and caused the glass at the end to glow. In the classic Crookes tube demonstration, the flow is from a negatively charged cathode, down the tube, past a positively charged anode to hit the glass, with the anode's shape (often a Maltese cross, for some reason) left dark as a shadow.

As researchers became more familiar with these 'cathode rays', they painted the end of the tube with a substance such as zinc sulfide, which was more fluorescent than glass, producing a brighter glow. The CRT (cathode ray tube) TVs were simply more sophisticated versions of this, where the 'cathode rays' were steered by magnets and electrical fields to create a picture. But what were these rays? Some thought they were charged matter, atoms that had picked up an electrical charge (what we'd now call ions), while others thought they were a different form of electromagnetic radiation.

The Cambridge-based physicist J.J. Thomson managed to measure the mass of the particles that made up this ray, showing that they certainly weren't a form of light, using a combination of the heat they produced when hitting a metal junction and the amount they were deflected by magnetism. However, his result suggested they weren't ions either. Thomson found that the ray's components were at least 1,000 times lighter than the smallest atom, hydrogen, and later refined this to around 1,800 times lighter than hydrogen.

These particles, which seemed to be coming out of stray atoms of gas or the matter of the electrodes – and so were emerging out of atoms – were much smaller than the atoms themselves. It appeared that the individual atoms were being cut – or at least that tiny parts of them were being pulled off. These particles, which Thomson called corpuscles, were soon better known as electrons, a name that had already been used as the basic unit of electrical charge produced by a battery.

Of itself, the existence of electrons did not say too much about what was going on inside an atom. Just because something comes out of a black box, that doesn't mean we know what's going on inside. Thomson's own theory, often called the plum pudding model, was that atoms were made up of a collection of negatively charged electrons, scattered through a massless, positively charged 'matrix' that held them in place electrostatically, making the electrons the 'plums'* in the model.

One aspect of Thomson's model that now seems strange is the idea that this positive matrix had no mass. This meant that if Thomson's model was accurate, all the mass of the atom had to come from its component electrons. With modern figures for the mass of atoms, this would have meant that hydrogen, an atom we now know contains a single electron, would have required a total of 1,837 electrons in in order for it to have sufficient mass.

A more useful picture of the atom would emerge from Manchester, where the New Zealand-born physicist Ernest Rutherford had his lab. Rutherford's team were experimenting on alpha particles – positively charged particles with a mass closer to that of an atom than that of an electron, and which were emitted by newly discovered radioactive materials. Before he had arrived in Manchester, Rutherford had discovered two different types of rays, which he named alpha and beta rays, later renamed as particles. Each was electrically charged, but with opposite values, curving in different directions if passed through electrical fields. Alpha particles would later be identified as the nuclei of helium atoms, each

---

* Strictly speaking, the electrons are not represented by plums, but by raisins. Plum pudding was a distinctly misleading name for what is now called Christmas pudding.

particle consisting of two protons and two neutrons (the particles that make up the atomic nucleus), but this was not known at the time. Similarly, beta particles turned out to be high-energy electrons.

After a number of preliminary experiments, in 1913 Rutherford's team had set up an experiment whereby a stream of alpha particles from a radioactive source (radon gas) was directed towards a thin piece of gold foil. The experiment took place in a cylinder from which the air had been removed. An observer – usually either Hans Geiger* or Ernest Marsden, Rutherford's assistants, had to sit in the dark in low light conditions until their eyes had acclimatised and then peer through a microscope towards the gold foil. The end of the microscope had a zinc sulfide panel fixed to it, so any alpha particles heading in that direction would cause a tiny flash. The observer would watch at different angles, rotating the microscope between observations to detect how many, or indeed if any, particles were deflected.

The Manchester physicists expected that when the alpha particles came close to the atoms in the gold, some would be slightly deflected by interacting with the electrical charges in the atom – which was the case. But a totally unexpected result was that some of the particles bounced back near to the direction they came from. In one of his more famous quotes, Rutherford commented that 'it was as if you fired a fifteen-inch shell at a piece of tissue paper and it came back and hit you'. This could only have occurred if the positive charge in the gold atoms, rather than being the diffuse matrix

---

* The one with the counter.

that Thomson imagined, was concentrated into a small, dense core containing much of the atom's mass.*

It says something for Rutherford's team that they took the precaution of looking for alpha particles travelling in a totally unexpected direction. A less capable experimenter might well have failed to do this and would have missed the breakthrough observation. In any case, Rutherford stole a piece of terminology from biology and called this the nucleus of the atom.

The solar system model

The Manchester discovery led to the beguiling idea that atoms were similar to miniature versions of the solar system with the nucleus at the centre of the atom taking the role of the Sun and the electrons, flying around the outside, playing the parts of the planets. There was a pleasing sense of symmetry if this were the case – and there's nothing physicists like more than symmetry. However, there was a fundamental problem with this model of the atom, which was never taken seriously by the physics community, despite still tending to be used to this day as the graphic art representation of an atom.

There's one big difference between a solar system and an atom. The force between the central massive body of the star and the orbiting planets in the solar system is gravity – but

---

* Rutherford had already predicted some kind of central small charge in 1911 after less sophisticated experiments, but it was the 1913 version that cemented his theory.

in an atom, the force between the nucleus and the electrons is electromagnetic. And these two forces of attraction don't work the same way. In simple terms, you can't keep something in orbit using electrical charge.

It was already known by Rutherford's time from the work of the great Victorian Scottish physicist James Clerk Maxwell that whenever you accelerate an electrical charge, it gives off energy in the form of electromagnetic radiation. This is how radio works, for instance. The signal from the transmitter accelerates electrons up and down the aerial, losing some of the electrons' energy in the form of photons – electromagnetic radiation. However, to stay in orbit around the atom's nucleus, electrons would be constantly accelerating.

It doesn't seem that this is the case, as an orbiting body usually moves at a constant speed. However, there is still acceleration, because this is any change of *velocity*, which is a measure of both speed and direction of travel. To keep in orbit, the direction of movement is always changing. That's fine with gravity in charge, but it would mean that the orbiting electrons would quickly lose their energy in a flare of electromagnetic radiation and plunge into the nucleus. Every atom would collapse.

To solve the mystery of the atom, other physicists employed the new and rapidly developing quantum theory – of which more later, as it will be essential to explore some of the stranger properties of graphene. The quantum model of the atom began with the young Dane Niels Bohr, who, rather neatly, started work on the problem while spending a year working with Rutherford in Manchester. All we need for the moment, though, in our understanding of how atoms interact to form material structures, is the idea that atoms have a small, dense, positively charged nucleus and are surrounded in some way by one or more moving electrons, plus the additional observation of the relationship between atomic structure and the chemists' periodic table.

The idea of putting the different elements into a structured table had been around for some time before the Russian Dimitri Mendeleev produced his earliest attempt in 1869 at what we now know as the periodic table, grouping elements into columns based on similar behaviour, with increasing atomic weight as you move down the table. As the structure of atoms became better known, it seemed clear that electrons occupied one or more 'shells'* around the atom, each of which could only have a certain number of electrons in it. How they occupied it without atomic collapse was not yet clear.

The chemical behaviour of an element, it was discovered, crucially depended on the number of electrons, and the number of open spaces, in the outermost occupied shell. You may remember something from school science called 'valence' which describes how an element is likely to combine with other elements to form compounds – this is a direct reflection of the state of that shell. So it is the detail

---

* Shells as opposed to orbits, to emphasise that the solar system model doesn't work. The nature of shells requires quantum theory, but for now the idea is essentially that each shell is a bit like a track. Electrons can only run on these tracks or jump between them as a result of a quantum leap. The electrons can be configured differently within each shell – these different configurations are confusingly referred to as orbitals, though each possible orbital is a probability distribution – a mathematical description of the chances of finding an electron at a particular location – not in any sense an orbit like that of a satellite.

of the atomic structure which gives chemists and physicists, whether dealing with multiple versions of the same element or combinations of different elements, a mechanism to describe how they will interact with each other.

Bonding sessions

Whether they were supporters of the four/five element theory or atomists, early natural philosophers had recognised that something must enable different elements or atoms to stick together so that they could make up the more complex stuff that we experience all around us (not to mention the complexities of our own bodies). Even the simple model of wood being made of earth, air, fire and water (see page 21) required this. By Newton's time, there was speculation that these connections between the elements were physical links, perhaps due to the shapes of atoms (for those who believed in atoms), or the result of some type of inter-elemental glue. But Newton, with the success of his theory of gravity, and familiar with the effects of magnetism, preferred the idea that there was some kind of attractive force linking the component parts together.

Inspired by Rutherford's model of the atom and gradual realisation of the role of electron shells, the American chemist Gilbert Lewis had by 1916 come up with the idea of an electron pair bond, or covalent bond, linking atoms. This was where one or more electrons in an outer shell, instead of belonging to a single atom, were shared between two atoms, forming a bond between those atoms due to the electromagnetic attraction between the electron and the two atoms' nuclei. Effectively, each atom made a claim on this electron or electrons as part of its structure. The electrons were Newton's glue.

The same year, German physicist Walther Kossel came up with a different way that atoms could bond where the outer shell had one more or one fewer electron than the normal atom. The result would be that the 'ion'* with the extra electron would be negatively charged and the ion that was missing an electron would be positive – so the two could be electromagnetically attracted together in an 'ionic' bond. Once again, Newton had had the right idea with the concept of attraction.

It's the existence of these two types of bond that not only makes it possible for chemical compounds to exist – from the ionic bond forming the simple sodium chloride that is common salt to the vast array of covalent bonds in the huge molecules of DNA – but, even more fundamentally, allows nature to go beyond individual atoms or molecules to produce physical objects made up of many billions of atoms, all linked together by different kinds of bonds. It's thanks to bonds that we are able to have solid materials. And as we come to look at graphene's capabilities, the nature of its bonds will be an important part of the way it performs.

---

* The word 'ion' comes from the Greek present participle of 'to go' (i.e. going), reflecting its first use to describe whatever it was that went from one electrode to another during the process of electrolysis.

Solids and structures

It is something of a self-evident fact that not every solid is the same, even when made out of the same type, or types, of atom. The way the bonds form between atoms, and the structure that the arrangement of bonds produces, helps determine the substance's physical properties – not just how it looks but how it reacts with other substances, its melting point, its strength and far more. As we shall see, it is the particularly impressive structure that carbon is capable of forming in graphite that results in graphene's remarkable properties.

Broadly speaking, solid substances tend to be either crystalline or amorphous. As we have seen, a crystalline substance is any one that has its atoms or molecules bonded together in a regular, recurring lattice. Many solids, from salt crystals to metals, do have such structures, but in others the bonds are higgledy-piggledy, without any repeating structure – these are amorphous solids, such as glass and many of the plastics.

However, while some atoms or molecules always form solids with the same kinds of structure, others have a range of options available to them – few more so than carbon – and the shapes formed by the bonds linking the molecules can have a crucial effect on the substance's physical characteristics, such as its strength, melting point and electrical conductivity.

A familiar example of the impact of the lattice shape on the physical characteristics of a substance is provided by solid water – or ice, as it is better known. Each individual water molecule does not have its three component atoms in a straight line, but instead has a wide-angled V-shape,* with the hydrogen atoms at the top points of the V and the oxygen at the bottom. This shape, combined with the attraction between the relatively negative oxygen atom in one molecule and the relatively positive hydrogen in another – an attraction known as hydrogen bonding – makes it fairly easy for the water molecules to form crystals surrounding a hexagonal space (and is responsible for the six-pointed nature of snowflakes).

Because of that particular shape and angle between the bonds, the lattice they form is not the most tightly packed that they can be. Water molecules can get closer together in a low-temperature liquid than they are when the solid crystal forms. This means that as water freezes it has the unusual (although not unique) property of becoming less dense as a solid than it was as a liquid. This means that ice floats on water (and tends to burst through containers that it is frozen in).** The simple hexagonal form of normal ice, incidentally, is by no means the only structure that water can adopt as a solid. There are at least seventeen different structures it *could* produce, but at usual freezing temperature and under the Earth's atmospheric pressure, the hexagonal form dominates.

---

* The angle is around 105 degrees.
** Ice's ability to float on water has helped shape the biology of fresh-water species. If it didn't, water would freeze from the bottom up and would not leave an insulated layer of liquid water beneath the ice for life to survive in.

Carbon is also able to solidify in a range of structures, which enable exactly the same element arranged in different configurations (known as allotropes) to behave as if they were unrelated substances. The best-known allotropes of carbon are diamond, which has an interlocking cube-shaped lattice of atoms, giving it great strength, and graphite, which, as we've already discovered, is made up of atom-thin layers of repeating hexagonal structures, individual layers being known as graphene. It's also possible to have carbon structures that make up relatively small molecules with closed structures, their bonds being like the pentagons and hexagons that make up the lines on the outside of a football.

These closed forms are known as fullerenes (or buckeyballs as a less serious nickname), both references to the American architect Buckminster Fuller, who designed geodesic domes that were similar to parts of a fullerene. The best-known buckeyball molecule, buckminsterfullerene, has 60 carbon atoms in its structure. Another, more open form of fullerine consists of a tube made of the same flat carbon lattice as graphene, wrapped around to form a cylinder. Such 'carbon nanotubes' are, in effect, tiny tubular pieces of graphene (think of taking a piece of paper and rolling it to produce a tube). The carbon fibres embedded in a polymer in everything from car dashboards to bicycle frames, producing a material that is misleadingly usually just called 'carbon fibre', may well contain some nanotubes, although they are mostly just strands of carbon chains, a bit like multiple thin strips of graphene. Although mostly artificial, fullerenes can occasionally occur in nature.

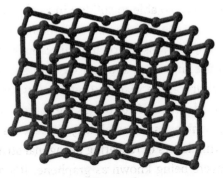

Diamond

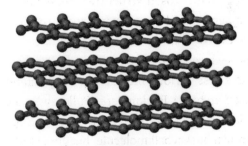

Graphite

Buckminsterfullerene

Two further allotropes of carbon are significant. One is lonsdaleite (named after the British crystallographer Kathleen Lonsdale), which is like diamond but has a hexagonal, graphite-like lattice instead of the usual cubic lattice. Lonsdaleite was first found in a meteorite and has also been artificially produced by putting graphite under high pressure and temperature. It should, in principle, be even harder than diamond, though existing specimens have tended to have a lot of impurities and more incomplete lattices than a good-quality diamond, making it weaker than its more familiar cousin. The other, less fancy allotrope is amorphous carbon, which lacks a uniform lattice structure. This is most familiar as coal or in the flecks of carbon that make up soot.

A moment of health and safety

Since we've brought up carbon nanotubes and slivers of graphene, it's worth bringing up a potential health hazard associated with them. While carbon itself is non-toxic and is sometimes prescribed medically to pick up unwanted material in the stomach, the tiny forms of carbon, if allowed to float around in the atmosphere, could give humans a similar problem to asbestosis. If the tiny fibres or ribbons are inhaled, they are small enough to cause damage inside the lungs, reducing the organs' effectiveness and increasing the risk of cancer.

In practice, in most of the applications of carbon nanotubes and graphene ribbons (a large sheet would be too big to cause a problem) the graphene is either embedded in a composite – as is the case with carbon fibre materials, with the carbon playing a similar role to glass fibres in fibreglass – or attached to a device, as we'll see in the various applications of graphene in later chapters. However, there is a potential risk when large-scale manufacture of graphene and carbon nanotube products is under way, and appropriate health and safety regulations need to be observed.

In this case, the important factor is the sheer thinness of the tubes or ribbons, which make it easy for them to get into the lungs. But equally important to the usefulness of carbon allotropes is the internal structure of the material.

Shapes rule

What makes the different structures of carbon interesting is that the way the carbon atoms are linked together has a huge influence on the physical properties of the material, such as strength, electrical conductivity and heat conductivity. Although diamond and graphite are made up of exactly the same atoms – carbon with six protons and six neutrons in the nucleus, plus six electrons* – by virtue of their different structures, they become radically different substances.

The most immediate difference is that diamond is transparent, while graphite is opaque. A transparent material allows light to pass through it. Some of the light may still

---

* It's these six electrons that make carbon so versatile and the backbone of the chemicals making up life on Earth. Carbon has four electrons in its outer shell and four vacancies, allowing for a wide range of structures, both when linking to other carbon atoms and in the many and extremely varied organic compounds necessary for life.

interact with the atoms: an electron can absorb a photon of light and jump up to a higher energy level, but then the electrons will soon re-emit another photon to continue on its journey through the material. Diamond has a good structure to allow this kind of passage. By contrast, the multiple sheets of graphite, which are aligned so that the atoms in one layer sit above the gaps in the next, succeed in blocking the passage of the photons entirely unless we have a very thin slice with only a small number of layers.

Equally, graphite, as we have seen, is very soft due to the ease with which the sheets of graphene pass over each other – but diamond is renowned for its hardness. Electrically, the allotropes are also distinct opposites: graphite is an excellent electrical conductor (and graphene, as we shall see, far more so), while diamond, though rarely employed this way for reasons of expense, is one of the best electrical insulators there is. Again, it is the crystalline structure that makes all the difference. In graphite's hexagonal structure, each carbon atom is connected to three others, leaving a loosely attached fourth electron from each of the atoms' outer shells able to float through the material and conduct electricity. Diamond, by contrast, has each carbon atom bonded to four other atoms, leaving no free electrons in the outer shells to conduct.

It's often the case that good electrical conductors are also good conductors of heat, and vice versa, because the same free electrons that carry electricity can be used to transmit heat energy. But diamond is something of an oddity in this respect, as it is an excellent conductor of heat – five times as good a heat conductor as copper. It can do this because it is also possible for heat to pass through a solid as vibrations, transmitted through the bonds between the atoms. Bear in mind that temperature is just a measure of the energy of the atoms that make up a substance. At a high temperature, the atoms jiggle around far more than they do at a low temperature.

The more rigid the bonds in a substance are, the less heat energy is lost as vibrations move through the material,* so a very rigid structure like that of diamond can make for a very good heat conductor. In fact, diamond is so good in this role that artificial, high-purity diamonds are the best known thermal conductors of any solid. Boring old carbon really is quite remarkable.

Going small

So far, what we've seen is comprehensible in terms of classical physics, the kind of physics that was understood in the 19th century and is still largely what we are taught in school. But to really grasp the significance of ultrathin materials such as graphene, we need now to take a plunge into the quantum world. Here small objects like atoms and electrons behave quite differently from regular objects we can see and touch – and it is these quantum properties that give graphene and other ultrathin substances many of their capabilities as wonder materials.

---

* You can see why a rigid substance has less loss of vibration by thinking of trying to send a pulse through a piece of cloth and something rigid like a pen. Push one end of the cloth and the movement is lost in the floppiness of the material, but push one end of a pen and the movement easily reaches the other end.

QUANTUM REALITY

3

Why quantum makes the difference

Some of the remarkable abilities of graphene that we will explore later are down to its flexibility and strength, which we can understand from its lattice structure using traditional Victorian physics. But to be able to explore its significance for the future of electronics, we need to have a basic grasp of quantum physics. This is one of the most essential aspects of physics, yet remains one that most of us know least about.

The word 'quantum' gets bandied around in all sorts of unlikely scenarios from the *Quantum Leap* TV show to Quantum dishwasher tablets and websites offering 'quantum healing'. When scientists use the word, though, they are thinking of something much more precise. A quantum is a separable chunk of something, and quantum physics reflects the aspects of nature that come in separable chunks, rather than having a continuous nature.

To take an everyday example, my local petrol station is currently selling petrol at 116.7 pence per litre. That's effectively treating price as if it were a continuum. They might equally sell it at 116.682314159 pence per litre if they really wanted to. But when it comes to *paying*, if I buy the minimum five litres, I can't pay, say, 583.5 pence, because the British cash system is quantised in units of 1p. There has been no such thing as half a penny since 1984, and there has never been a smaller unit of currency since decimalisation. So, if I bought exactly five litres, I would have to pay either 583p if the company were generous, or, more likely, 584p because they had rounded the value up to the nearest whole penny.

It turns out that a lot of aspects of nature that had once been thought of as a continuum until the early 20th century – a beam of light, for example – are in fact quantised, and so come in minimum-sized chunks or 'quanta'. In the case of light, these quanta are called photons. The name 'quanta' (plural of quantum) dates back to Max Planck, the German physicist who first considered light to be broken up this way. He didn't like the idea, because everyone at the time thought that light was a wave; looking back in his old age, he said: 'Briefly summarized, what I did can be described as simply an act of desperation.'

Planck assumed that quanta of light were just a helpful calculation tool, but Einstein would demonstrate that they must actually exist. It's not that the wave theory of light was entirely wrong – light often does act as if it were a wave, but there are times when it can only be understood if we consider it to be acting as a collection of quanta – photons. By this time, scientists were also already aware of the effective quantisation of matter, into atoms (or subatomic particles to go for the ultimate matter quanta, as far as we know). Even here, our senses can deceive us. It appears that a stream of water, or a piece of rock, is a continuous thing, but we know that in reality it is made up of tiny, separate particles, held together by bonds.

Of itself, the existence of these particles is not such a surprise. What shook the early 20th-century physicists was that the basis of every normal object was revealed to be a collection of particles which refused to behave the way that they were expected to. Quantum particles do not behave like everyday objects. Such was the resistance from Albert Einstein – who had been instrumental to proving the existence of the quantum world – that for years he would regularly come up with thought experiment challenges which he hoped would show that quantum theory was wrong. Every one of his challenges proved ineffective.

What offended such a great mind as Einstein? He famously wrote to his friend Max Born: 'The theory says a lot, but does not really bring us any closer to the secret of the "old one". I, at any rate, am convinced that He is not playing at dice.' The choice of imagery was a reaction to the way that quantum theory has probability at its heart. As we will see, crucially for some electronic applications, quantum particles that haven't interacted with something else for a while aren't situated in one place, but rather exist merely as a collection of probabilities for different locations. Such uncertainty made Einstein also write to Born on another occasion: 'In that case, I would rather be a cobbler, or even an employee in a gaming house, than a physicist.'

Three quantum essentials

Perhaps the best known of the probabilistic aspects of quantum theory, at least by name, is Heisenberg's uncertainty principle. This does not mean that 'everything is uncertain' as the term is sometimes loosely used to imply. It's quite the reverse, in that the uncertainty principle describes a series of precise relationships. It tells us that any quantum particle has pairs of properties associated with it where, the more accurately we know one value, the less accurately we can pin down the other. So, for instance, the more accurately we know the location of a quantum particle, the less accurately we can know its momentum. We can never know both perfectly at the same time. If we exactly know the momentum, say, the particle could literally be anywhere in the universe. Similarly, the more accurately we can pin down the energy of a quantum particle,* the less accurate we can be about the timeframe in which we make the measurement.

Even more fundamental to quantum behaviour is the Schrödinger equation. This describes the way that a quantum system – in its most useful simple application a single quantum particle – changes over time. When the equation was first developed it caused considerable confusion, as it was assumed that it dealt with the *location* of a particle. And if this were the case, it seemed to say that, over time, a quantum particle would spread out, occupying more and more space. This doesn't happen (thankfully). But the same Max

---

* Strictly this refers not to a particle but a quantum system, which could include many particles, or indeed empty space with no particles present.

Born to whom Einstein wrote about his quantum concerns realised that the equation* did not describe the location of a particle, but rather the *probability* of finding a particle in a particular location. Over time, the range of possible locations where the particle could be spread out through space.

This provided a radically different view to the then conventional conception of quantum particles such as electrons and atoms being like tiny balls which were able to move around but which were in a single, specific location at any one time, just like the kind of ball that we play with in a game. Instead, once a quantum particle has been created, its potential locations spread further and further out. It's not that it has equal probability of being in all those places. It is typically more likely to be where we would expect it to be if it were a conventional ball – but it also has a possibility of being in totally unexpected places, and in some cases even the most likely location is totally different from what we'd expect from experience. All that exists at this point, when the quantum particle is not interacting with anything, is a smear of probabilities across space. The particle is not, as is sometimes described, in two places at once. It does not have *any* location at all. It's only when the particle interacts with something else at one of its possible locations that its settles down on a specific place to be. We can calculate the exact probability of any particular location being the one where we will find the particle, but until it interacts with its surroundings we have no idea which one of the possible locations will prove correct.

---

* More precisely, the square of the equation.

This 'smeared out' nature of quantum particles enables a third quantum oddity which turns up regularly in electronics and will be crucial to using graphene: quantum tunnelling. In the conventional world, if I send an object towards a barrier, and the object hasn't enough energy to get over or through the barrier, it can't go any further. It stops. But by the time a quantum particle is most likely to be in the vicinity of a barrier, because the possible locations have spread out, there is a probability (usually small) that it is already the other side of the barrier. In the event that this turns out to be the case, it's as if the particle has tunnelled through the barrier and out the other side – taking no time in the process. It's already there. That is quantum tunnelling.

It's a bit like someone throwing a tennis ball repeatedly at a wall and finding out that every now and then the ball doesn't bounce off it and fall to the ground, but is already heading away from the other side of the wall. This sounds ridiculous – yet quantum tunnelling has been observed many times and is both frequently used and occasionally a problem to avoid in electronics. It's common, too, in nature. We wouldn't even exist without quantum tunnelling, as it is needed to make the Sun work.

The Sun produces the energy that has enabled life to thrive on Earth using a process known as nuclear fusion, where the nuclei of hydrogen atoms are joined together to produce the heavier element helium, giving off energy in the process. Inside a star like the Sun, hydrogen nuclei are squashed together under immense temperature and pressure – but it's not enough to get them to fuse. Because they are positively charged, the electrical repulsion between the nuclei prevents them from getting close enough for the fusion process to occur. It's only because the hydrogen nuclei are quantum particles and can tunnel through the barrier formed by the electromagnetic repulsion that the Sun can function.

In electronics, tunnelling is something that circuit designers have to be aware of. If they make parts of the circuit on a chip too close together, electrons can tunnel through the barrier between two parts of a circuit, resulting in a glitch in the system. On the positive side, though, tunnelling has proved a boon for constructing certain types of transistor and for keeping information when the power is turned off.

We now take it for granted that we have memory chips that function without power. This so-called flash memory is used in phones, memory sticks and the solid-state drives that are fitted in many modern computers instead of the old hard discs. Conventional computer memory simply loses the electrical charges that make up the 0s and 1s of the data held in it when the device is turned off. But flash memory hangs onto those charges despite a lack of electrical power. This is because each bit of the flash memory is stored in a tiny insulated island. It is only by intentionally getting electrons to tunnel through the barrier that the value in the memory can be changed.

When we think specifically of graphene and its curious properties, we need to get a quantum view of what's happening to the electrons that are part of the atom, for which we will have to go back to Manchester at the start of 1912, when the young Niels Bohr had just arrived at a new laboratory.

Bohr's atom

The 26-year-old Bohr had been given a grant by the Carlsberg* Foundation to spend a year studying in England. He had hoped to work with J.J. Thomson, the discoverer of the electron and the man behind the plum pudding model. But when Bohr turned up in Cambridge, equipped with an English copy of *The Pickwick Papers* and a Danish–English dictionary in an attempt to improve his English vocabulary, he rapidly found that Thomson had little interest in his work. This might not have been helped when Bohr, on their first meeting, took the chance to point out to Thomson some errors in the older man's recently published book.

After a few uncomfortable months at Cambridge, Bohr managed to get a transfer to Manchester, where he found the jovial, loud figure of Ernest Rutherford a much more amenable and effective mentor. Bohr himself was a quiet introvert, who struggled to put his thoughts into words, but he had a huge admiration for Rutherford and the way he worked so openly with the young physicists in his team. When Bohr had his own team, he very much based the way that he worked with them on Rutherford's example.

Bohr was set to work with alpha particles, at the peak of their interest in the Rutherford lab, but quickly found a greater enthusiasm for exploring the structure of the atom beyond the newly discovered nucleus. Strangely, Rutherford

---

* Yes, that Carlsberg. Later on, the Danish Academy of Science would give Bohr tenancy of the Aeresbolig or House of Honour, a mansion provided to the Danish nation by Carlsberg, which came with a lifetime supply of lager.

himself was not particularly concerned with the topic. He was more interested in the mechanism of scattering incoming particles by the atomic nucleus than exactly what was going on in the detailed structure of the atom. However, Bohr picked up on the work of Charles Galton Darwin (the grandson of the better-known Charles Darwin), who had suggested that alpha particles that passed near an atom without bouncing off the nucleus were being slowed down by interacting with the negatively charged electrons around it.

Bohr started to think about how the electrons around the atom managed to stay tied to the nucleus without plummeting into it. As we have already seen (page 32), it was not possible for them to be orbiting like satellites around a planet. But perhaps he could find some other way that they could stay in place but remain stable. He wrote to his brother Harald:* 'Perhaps I have found out a little about the structure of atoms. Don't talk about it to anyone … it has grown out of a little information I got from the absorption of alpha rays.'

Bohr knew that there was no stable way that electrons could either be arrayed stationary around the atomic nucleus or in conventional orbits. He had to come up with a more radical solution. Using the quantum idea that had been started by Max Planck and built on by Einstein (before he turned against it), Bohr suggested that electrons could

---

* Harald, two years younger than Niels, was a gifted mathematician with a talent for football, playing for Denmark in the 1908 Olympic Games. Niels, though not up to Harald's standard, also had some success on the football field, as a goalkeeper.

only inhabit particular orbits. Instead of moving incrementally from one orbit to another, as a spaceship would do, he believed that it was impossible for electrons to exist in between the orbits, so they made an instant jump from one to the next – a so-called quantum leap.* This way, the orbits themselves would be quantised.

The available orbits were linked to the energy of the electron. The approach made sense particularly if the electron was thought of as a wave. Where Planck and Einstein had shown that light, usually thought of as a wave, could behave as if it were a collection of particles, so electrons, usually thought of as particles, could also behave like waves. The energy of a quantum particle corresponds to the frequency of the wave. The higher the frequency (or the shorter the wavelength), the more energy the equivalent particle has.

When an electron made a quantum leap to the next-highest orbit, the amount of energy it gained was the equivalent of a wave getting an increase in frequency. But if the electron *did* act like a wave, it would have to pass around the atom in a full number of wavelengths – the waves had to match up when they met themselves having passed around the atom – and this meant that only certain wavelengths, hence specific energies of the electron, were allowed. If the electrons around an atom behaved as waves, the orbits had to be quantised.

What clinched it for Bohr was accidentally discovering the earlier work of a Swiss physicist, Jakob Balmer, who had

---

* Ironically, given the way it's used in ordinary English to imply a major change, a quantum leap is actually the smallest possible change that an electron can make.

produced a formula that predicted the spectral lines of hydrogen. When an element is heated, it does not give off every colour of light, but rather produces a set of separate, specific colours (frequencies). Balmer's equation matched Bohr's idea for how the electron orbiting a hydrogen atom could be allowed to jump from orbit to orbit. The gap between two orbits would be a fixed amount of energy; and the colour of the light given off when an electron jumped down across that gap – which corresponded to the energy of the photon produced – matched the spectral frequencies that Balmer's theory predicted. It could surely not be a coincidence.

Although we have been calling these possible levels the electron could occupy 'orbits', it wasn't really an appropriate term to use, as the electrons were restricted to specific options. It was more as if they were running on rails that surrounded the nucleus, rather than behaving like an orbiting satellite. Bohr called these allowed energy levels 'stationary states', with the lowest possible energy called the ground state.

Bohr's model only worked for hydrogen. It would take the full-scale quantum theory that was developed a decade later to get a better picture that applied to all the elements. In essence, though, Bohr's stationary states were the shells around the atom that electrons can occupy. Within these shells, given the Schrödinger equation's prediction that an electron should exist as a cloud of probability rather than a classically orbiting body, the *orbitals* are the different possible probability distributions which begin with a simple spherical shell, but rapidly develop more complex lobed shapes as higher energy levels are reached.

From orbitals to band gaps

In a solid that is going to be used in an electronic device, the possible orbitals around the different atoms in the solid can and do interact as the orbitals overlap. This results in multiple possible orbitals within the material for each orbital that a single atom of that material could have. In fact, if there are 1,000 atoms in a lattice like that of graphene, each carbon atom has 1,000 different possible orbitals for each of its original ones.* In practice, there will usually be many billions of atoms, and so billions of tightly packed orbitals, so close together that they can be considered continuous bands, and are referred to as such.

The different structures that the atoms can take mean that rather than having a continuous band featuring all possible values, there is often a pair of particularly significant bands with a gap between them. This is known as the band gap. If the outer electrons are within the bottom band, known as the 'valence' band, they are tied to the atom and tend to be involved in forming bonds. If they are in the upper band, the 'conduction band', their attachment to the atom is sufficiently weak that they can float through the substance and conduct electricity.

In an insulator, all the outer electrons remain within the valence band and never have enough energy to cross the gap and get to the conduction band. A semiconductor, as used in electronics, still has a band gap, but it is small enough for a reasonable number of electrons to cross it. Conductors either

* This only applies to the electrons in the outer shell, the so-called 'valence' electrons that are involved in forming bonds. There is very little overlap with inner electrons, so their bands are negligible.

have a very narrow band gap or none at all – and typically already have electrons in the conduction band. Graphene has a 'zero band gap' – the valence and conduction bands line up exactly with no gap or overlap. The valence band is well occupied, but there is nothing yet in the conduction band. Even so, this already makes graphene a good conductor.

Of itself, though, we haven't quite got enough quantum theory to explain why graphene has such *extraordinary* abilities as a conductor. For that we need to go beyond the Schrödinger equation to the Dirac equation.

Dirac's contribution

Bristol-born Paul Dirac is probably the least well-known name among the greats of quantum theory. Most have at least heard the names Heisenberg or Schrödinger, but Dirac's name would draw a blank – even though the contribution he made to quantum physics was just as important. In part this is probably because unlike, for example, the outgoing Einstein, Dirac was pathologically shy and given to making remarks that did anything but put other people at ease. Famously, Dirac was once giving a lecture while visiting Wisconsin. After rattling through his material at high speed, he asked for questions. An audience member said to Dirac: 'I don't understand the equation in the top right-hand corner of the blackboard.' Dirac simply stared ahead without replying as if he had not heard the question. After an uncomfortable silence, Dirac was asked if he had an answer and retorted: 'That was not a question, it was a comment.' Some wits might have come up with this response with intended humour, but Dirac was deadly serious.

Although, like many scientists, Dirac did make such occasional forays out into the world to visit other institutions, and enjoyed a walk out into the countryside every Sunday to clear his mind, he was most comfortable in his Cambridge study, where his laboratory equipment (like Einstein's) was a pencil and sheets of paper. One of his early contributions to quantum physics was to show that the approach taken in Schrödinger's equation was entirely compatible with earlier work that Heisenberg had done. Heisenberg's version of quantum theory, matrix mechanics, worked purely by manipulating arrays of numbers, without any real-world model to suggest what was involved. Some loved its mathematical purity, but others felt it was difficult to understand with its lack of connection to anything that could be envisaged. Dirac bridged the two. However, his biggest success would be in taking Schrödinger's equation to the next level.

Schrödinger's elegant piece of mathematics comes in two forms: the more complex time-dependent Schrödinger equation, which is the one that shows the probability of a particle's location spreading out over time; and the time-independent equation, which describes the behaviour of a quantum particle that is in a 'stationary state', such as an atomic orbital. These equations are very effective at describing what happens to quantum particles, but they have one severe limitation. They are known as 'classical' equations, meaning that they assume Newton's laws of motion, rather than the more sophisticated variation that has to be introduced when using Einstein's special theory of relativity.

If a particle is moving slowly, this isn't a problem. Special relativity only makes a significant difference when something is travelling at a reasonably large fraction of the speed of light. However, there are circumstances when quantum particles do move extremely quickly – including in the case of electrons in an atom or the charge carriers in graphene. Dirac felt it should be possible to combine the type of information that comes out of Schrödinger's equation with the impact that the special theory of relativity has on motion.

After a frantic period of work leading up to Christmas 1927, Dirac came up with his own equation for the behaviour of the electron. The equation was in four parts, which not only incorporated the special theory of relativity, while collapsing to the Schrödinger form at low speeds, but also handled another aspect of the behaviour of quantum particles called spin,* which had yet to be properly dealt with by existing mathematics.

When his work was published by the Royal Society in February 1928, Dirac's conclusions proved a significant shock to the physics world. It wasn't so much the mathematics he used, though the work was anything but simple in that respect, but rather the implications of his equation. It would not work unless electrons could have either positive or negative energy – and the very concept of a negative amount of

---

* As is often the case with quantum theory, things are not what they seem when it comes to spin. This property of quantum particles has nothing to do with rotation, and when measured along any chosen axis can only have one of two values, up or down. But it shares some similarities with the property of large-scale objects called angular momentum, and so spin was used as a name, even though it is a distinctly misleading one.

energy was a baffling one. Worse still, the implication was that an electron could not just take quantum leaps down to a lowest positive level, the ground state, but would continue to jump down below zero with an infinite possibility of extra leaps. Yet they clearly didn't do this.

The Dirac sea

For a while it seemed likely that the possible negative energy solutions to the equation would simply be ignored. There was some precedent for this. When James Clerk Maxwell had come up with his equations for electromagnetism, which had shown that light was an interaction between electricity and magnetism, the equation describing an electromagnetic wave had two distinct solutions. One was for the wave that was well known in nature, travelling from transmitter to receiver at the speed of light. But there was also a solution where a form of light wave left the receiver at the time of arrival of the normal wave and travelled backwards in time to arrive at the transmitter at the time the normal wave departed.

Both these solutions to Maxwell's equations were equally valid – and the backward travelling wave would eventually be useful.* However, it was usually the case that only the wave that travelled forward in time would be used and the other was ignored – swept under the carpet. After

---

* Richard Feynman and his senior John Wheeler used the hypothetical backwards-in-time wave to deal with a problem whereby an electron seemed like it should be influenced by its own electrical field, causing problems for the mathematics.

all, the forward-travelling solution perfectly matched what was observed, so why make things overly complicated? Similarly, with Dirac's equation, the positive energy solution did a wonderful job of matching observation, and many were happy to ignore the negative energy solution. But not Dirac himself.

Dirac would spend a good year battling the negative energy problem. He did not manage to remove it entirely, but rather came up with a scenario that meant it could exist while still usually being ignored. However, this scenario took a bit of getting used to. He imagined that every single negative energy level that an electron could occupy was already full up of electrons. This meant that the universe had a kind of infinite sea of negative energy electron positions, each occupied by an electron. Then the 'real' electrons that we observe would have to have positive energy, because all the negative levels were already filled, and a law of physics called the Pauli exclusion principle required that no two electrons could have exactly the same properties, including their energy level.*

Although this scenario seemed more than a little unlikely, it did make a prediction – and one that could be tested – that made it different from a situation where negative energy just didn't exist. Inevitably, sometimes one or more of the negative energy electrons would be hit by a photon and would

---

* This may seem to contradict the idea we've already seen that an atom can have a number of electrons in the same shell. However, although these electrons occupy the same energy level, they must have other properties such as their spin with different values. The Dirac sea deals with all possible negative energy electrons.

jump up in energy to a positive level, just as electrons jump between the normal positive energy levels. This would leave holes in the negative energy sea where the negative energy electrons had been. When that happened, ordinary, positive energy electrons could drop down into the holes, disappearing from the normal positive energy world while giving off photons. So, experimenters could look out for these negative energy electron holes, or rather their impact. Having a hole – in effect, an absent, negatively charged, negative energy electron – turned out to be identical to having a present, positively charged, positive energy electron. So, Dirac's theory predicted that there would be a particle that was exactly like an electron, but with a positive charge.

If such a positively charged particle were found and it met up with a normal electron, it would be like a normal electron dropping into the hole in the sea. Both the positively charged particle and the electron would disappear, giving off electromagnetic energy in the form of a pair of photons. The positive particle, soon to be called a positron, or an anti-electron, would be discovered in 1931 by Carl Anderson, an American PhD student, in cosmic ray showers, when high-energy particles from space crash into the Earth's atmosphere. Ironically, when a lecture was given at Cambridge about the discovery of positrons, Dirac happened to be out of the country and didn't hear about it until some time later.

Alternative ways to approach Dirac's equation were later found, without the requirement for the infinite negative energy sea model, but the basic outcomes have stood the test of time. And because the electrical charge-carrying particles in graphene are travelling extremely quickly, they can only be effectively described using Dirac's equation rather than Schrödinger's. And as we shall later discover, this gives graphene remarkable electrical properties.

Quantum theory gives us all we need to comprehend what is happening inside a piece of graphene – and the same principles are essential for us to be able to produce the microelectronics that are found inside every computer, phone and electronic device, enabling quantum physics-based products to represent around 35 per cent of GDP in developed countries. However, there's one other logical requirement for us to understand the way that quantum physics enables us to make solid state circuitry and how ultrathin substances such as graphene can be involved in these devices. We need an idea of what basic electronic devices *do*, and how these are put together to make a fundamental logical concept called a gate.

Electronic components

Almost all electronic mechanisms depend primarily on two relatively simple components, the diode and the transistor. There are various other components, such as capacitors and resistors, but most of the functional parts of a transistorised device are based on these related components. A diode is simply a one-way path. It is an electronic component which allows electrical current to flow in one direction, but not in the other. There are a number of ways of making a diode, but the simplest form is a sandwich of two different types of semiconductor (typically materials such as silicon or germanium). A semiconductor, as we have seen, has a band gap, but it is one that can be bridged, sometimes by a secondary electrical current, sometimes with another source of energy, such as light.

One side of the simple diode is called a 'p-type' semiconductor. This has been 'doped'* with another material such as boron, which results in it having more gaps in its valence band than is normal in the semiconductor. These gaps, known as 'holes', rather like the holes in the Dirac sea (though these are positive energy holes) act as if they are positively charged particles.

The other side of the diode is an 'n-type' semiconductor. This also has been doped, but with a different material, such as phosphorous. An n-type semiconductor has relatively few holes in its valence band, but a lot more free electrons in its conduction band. When the diode is connected into a circuit, if the n-type side is on the negative side of the circuit and the p-type side is on the positive side, electrons will flow through the diode, attracted by the positive holes in the p-type side. However, if the circuit is connected the other way, the excess electrons on the n-side repel any further electrons, so current won't flow.

The simplest form of transistor is not dissimilar to a diode but has an extra connection to a central piece of material that is sandwiched between the two outer sections. In such a transistor, the layers are typically either n-type/p-type/n-type or p-type/n-type/p-type. With this type of set-up, changing the voltage that is applied to the

---

* Doping a semiconductor is intentionally introducing impurities into it.

middle section of the sandwich, called the base, enables the transistor to act either as an amplifier or as a switch.

When the transistor is in amplifier mode, small changes in the voltage applied to the base result in much bigger changes to the voltage across the two outer sections. When it acts as a switch, the difference between having a voltage on the base and having no voltage enables the transistor to switch off and on a current that is trying to flow through the outer parts.

In practice, modern circuits usually make use of a different type of transistor, called a field effect transistor. Here, instead of a central base, the switching part of the transistor, called a gate, is separated from the other parts by a thin insulating layer. In this type of transistor, it is the electrical field generated by the gate that allows it to control the flow through the device. In the case of graphene, as we shall see, this field effect is very pronounced, making it good for producing amplifying field effect transistors. However, the difficulty of getting it to stop conducting means that, alone, graphene isn't suitable for making a switching transistor (though as we shall also see, there are a number of ways around this).

The reason the ability of a transistor to have a switch action is important is that switching is an essential aspect of the fundamental unit of computer hardware: the gate.

Jumping the gate

In physical terms, a computer chip usually contains a complex circuit, typically built up on a silicon wafer base – but in logical terms it is made up of gates. These are parts of the circuit that represent a logical operation, usually described using the terms of Boolean algebra. Named after the 19th-century English mathematician George Boole, Boolean algebra uses a relatively simple structure to combine true and false statements, and is fundamental in computing.

The key requirement to make a computer function is to be able to deal with the binary 0s and 1s that represent numbers and instructions bit by bit. (A 'bit' is just a 'binary digit'.) While Boolean algebra was originally designed, well before the existence of the first electronic computers, to deal with problems involving 'true' and 'false', it proved equally effective in manipulating 0s and 1s – in both cases it's a matter of dealing with a system that can have only one of two values. We can think of 0 as false and 1 as true.

Each type of logic gate in a computer manipulates numbers in different ways. Most combine two different inputs to produce one output, but the simplest gate merely transforms a single value to the opposite one. This is the NOT gate, which flips the bit. If the value is currently 0 it becomes 1; alternatively, if it's 1, it becomes 0.

Moving on to the gates that combine two inputs, we start with AND and OR. The AND gate produces 0 in every possible combination of its two inputs (let's call them A and B), except when both inputs are 1. If we think of 0 as false and 1 as true, the AND gate produces true only if both input A is true AND input B is true. You can think of this in logical terms by saying of a vehicle 'This is a red bus.' The statement is true only if the vehicle is red AND it is a bus. If it is just red but not a bus, or a bus but not red (or neither a bus, nor red) it is not a red bus.

The OR gate, by contrast, is less fussy in the way it operates. It will produce a 1 if either input A is 1 OR input B is 1 – and will also do so if both are 1. The only circumstance where it will produce a 0 is if *both* A and B are 0. In logical terms, it's like looking for an object that is either red, or is a bus. A red postbox would be a true match, and either a green or red bus would also be true. But a yellow postbox would not match the criteria and so is false.

Each of the AND and the OR gates have negative alternatives, known as NAND and NOR. These produce exactly the same effect as putting the output of an AND or an OR gate through a NOT gate. So, where the AND gate produced 1 only if both A and B were 1, the NAND gate produces 1 *unless* both A and B are 1. Similarly, where the OR gate produces 1 except when both A and B are 0, the NOR gate produces 1 only if both A and B are 0.

Finally, we have a subtly different kind of OR – the XOR gate, which stands for 'exclusive OR'. If you remember, the OR gate produces 1 if A is 1, if B is 1 and if both A and B are 1. The XOR gate, as the name suggests, requires an exclusive selection. It produces 1 as output if A is 1 or if B is 1 – but not if both are 1. A and B must be different. So, if both A and B are 1, it produces 0. This would be the equivalent of looking for something that was red, or a bus, but not a red bus. (There is also a negative version of this, the XNOR gate, which produces 1 if both A and B are 0 or both A and B are 1. It produces 1 when the inputs have the same value.)

Technically, there really is not a need for more than one kind of gate – we only need the electronic structure to produce a NOR gate or a NAND gate. Each of these, linked with more of the same kind, can produce all the other types of gate. For example, if you wanted a NOT gate, you could join together both inputs to a NOR gate. This would mean that both A and B would always have the same value. A NOR gate receiving two 0 values produces 1, while with two 1 values it produces 0. By linking the inputs, a NOR gate is forced to act as a NOT gate.

Gates make up both computer memory and the processors that do the hard work. The transistors in the circuit are arranged so that they make up different kinds of gate. For example, a NOT gate can be made from two transistors, while a more sophisticated gate like a NAND gate can require as many as four transistors. Before integrated circuits, printed circuit boards would be made up combining thousands of individual transistors in this way. In a modern computer, though, the whole memory unit or processor is combined on a single chip with layers of semiconductors, insulators and conductors replacing the individual components.

To get graphene and other ultrathin materials to be able to produce the same kind of circuits – though in a much thinner and more flexible fashion, as we will see – it has been necessary both to think through the way to make these gates using the new materials and to deal with the pros and cons of the properties of the different materials – for example, graphene's extremely high conductivity.

Ultrathin electronics were to pose a unique and intriguing set of challenges. But first we had to be able to make the materials – the graphene – which for so long had been considered impossible to manufacture.

Which takes us back to Andre Geim and Konstantin Novoselov.

LIKE NOTHING WE'VE SEEN BEFORE

4

The road to Manchester

We have seen how Geim and Novoselov both moved to Manchester from the Netherlands. Geim had chosen the university because he preferred the British system and was offered a long-term post there. The fact that Manchester had offered his wife, Irina Grigorieva, a job as well, made the move more attractive than any alternative. Although Grigorieva had been a postdoc at Bristol, she had only found a role as a part-time teaching laboratory assistant at Nijmegen. But the board at Manchester were familiar with her Bristol work and she is now a well-established physicist in her own right, still based in Manchester. As for Geim and Novoselov, the pair may have first worked together in the Netherlands, but each had made a significant journey from his Russian origins.

Andre Geim was born in the Russian resort of Sochi on the Black Sea, near the border with Georgia. Now best known as the location of the 2014 Winter Olympics, in 1958, when Geim was born, Sochi was part of the USSR, a very different place from modern Russia, let alone the western European cities where he would later flourish. Geim spent his first few years living with his grandparents, as much of his family was incarcerated in the Gulag. His family were of German background and hence were considered potential enemies of the state in a part of the world still finding its feet little more than a decade after the end of the Second World War. Even though Stalin had implemented his own reign of terror, the dark hand of Germany was not forgotten.

Science had always been Geim's passion – as a boy, he won a regional Chemistry Olympiad by memorising a 1,000-page dictionary, and proved as excellent at the experimental side as the theoretical. With glowing results from his school and a perfect score in the exams he took at age sixteen, he had looked forward to studying physics at the Moscow Engineering and Physics Institute. Unfortunately, despite his excellent qualifications, he was rejected. Although there is no direct evidence, Geim believes to this day that this was due to his German family background. The teenage Andre spent some time working at the same engineering factory as his father to pay for extra tutoring in maths and physics in order to give himself even more of an edge – only to be rejected by the Institute a second time.

Luckily, there was not the same level of discrimination on the selection board of the prestigious Moscow Institute of Physics and Technology, generally known in the USSR as PhysTech. In some ways, it's a surprise that Geim did not apply to PhysTech in the first place. Set up at the end of the Second World War by leading Soviet scientists, the founding idea was to move away from the mass teaching methods used elsewhere in the USSR. Each of the students selected to attend would be given an individual programme of education, tailored to them and provided by leading figures in the field. This vision of a wholly independent institution foundered, as some of the scientists involved had been critical of the Soviet system, but they managed instead to set up PhysTech as a part of Moscow State University, where it was allowed a surprising degree of autonomy. Though Geim is rightly critical of the Soviet state, this institution's approach seems to have benefited him.

As was common with PhysTech students, Geim went on to join a section of the Russian Academy of Sciences, in his case, the Institute of Solid State Physics, where he gained his doctorate. After this, Geim worked mostly in the West at universities in Nottingham (where he first realised just how many obstacles faced those doing science in the Soviet system), Bath and Copenhagen, becoming an associate professor at Radboud University Nijmegen in the Netherlands in 1994. His move to Manchester in 2001, he claims, was in part due to the hierarchical, backbiting nature of the Dutch academic system, which he found less constructive than that of British universities.

In Nijmegen, as we have seen, one of Geim's doctoral students would be Konstantin Novoselov. Sixteen years younger than Geim, Novoselov was born on the eastern side of Russia in Nizhny Tagil, an industrial city where railway and military engineering dominated. While Novoselov was not initially the same kind of standout student as Geim, he exhibited an unusual level of curiosity about electricity and magnetism. Given a rather smart German train set at the age of eight, he was more excited by the DC controller for the set than the actual trains. With this he had a variable power supply, which he used to experiment over the years with electromagnets and electrolysis.

Like Geim, Novoselov won a place at PhysTech, still a significant force in Russian physics, from where he went straight to work under Geim in the Netherlands. The two hit it off, not just from having a common background, but in their approach to science. It seemed only natural that Novoselov should travel with Geim to Manchester, putting in place the second essential piece in the creative game that would result in the creation of graphene. (He continued to be officially registered in the Netherlands until 2004, when his doctorate was awarded there.)

From discovery to Nobel

Although there is no doubt that there was a sudden and significant mental shift when the Scotch tape method was dreamed up and proved successful, this doesn't mean that Geim and Novoselov went straight from their early Friday night experiments to winning the Nobel Prize. The initial breakthrough was followed by many months of solid work.

Novoselov, looking back on the first year that followed the 2003 discovery, has described it as 'a whole year of continuous excitement'. In real life (as opposed to TV and movies) science usually has many long periods with nothing much happening. But during that year, the pace was intense.

Novoselov again: 'For a typical piece of work, novel results and experiments come maybe on a weekly or daily basis. At that time, it was on an hourly basis.'

Graphene opened up so many possibilities that there was constantly something new to be investigated. And after the Manchester lab published their first paper in 2004, interest worldwide in this wonder material shot through the roof. It was only a matter of time before the two Mancunian Russians were awarded the Nobel Prize for Physics, winning it in 2010 for 'groundbreaking experiments regarding the two-dimensional material graphene.'

Referring to graphene as a two-dimensional material might seem an exaggerated boast. Any substance does, of course, have *some* depth, even if it is measured as a fraction of a nanometre. It is part of a three-dimensional world. However, while graphene may not be two-dimensional in the pure mathematical sense, there is no way to make anything thinner – it is as thin as you can get, the next stage in removing a slice being no atoms at all. This means that it shares some properties with theoretical two-dimensional objects and behaves in ways that its three-dimensional counterpart, graphite, can't.

The Nobel Prize has come to be recognised as the ultimate mark of excellence in the disciplines it covers, though it's worth remembering that many of history's greatest scientists have not won Nobels. The Physics prize is limited to a maximum of three people, who must be alive at the time of the award. It was first given out in 1901 to the German physicist William Röntgen 'in recognition of the extraordinary services he has rendered by the discovery of the remarkable rays subsequently named after him'. As it happens, the Nobel committee got one thing here wrong – the suggestion that the 'remarkable rays' would be named after their discoverer. The term 'Röntgen rays' was a non-starter as Röntgen's original description of these then-mysterious rays as 'X-Strahlen' or X-rays proved much more popular. Effects in physics are often named after the discoverer – but not fundamental phenomena.

Given the date of the first award in 1901, great names such as Galileo, Newton or Maxwell of course never featured. And it is worth remembering that the selection process for the Nobels is a very human one, which has resulted in some distinct oddities, notably the 1912 prize, won by the Swedish scientist Gustaf Dalén for the dubious honour of having invented a better gas regulator for lighthouses, just at the time when lighthouses were converting to electricity.

The process of nomination for the prize is not secret, even though the detail of who was nominated is kept under wraps for 50 years, so we do not know, for example, whether the Manchester pair were nominated before the 2010 prize – or how many nominations they received. It's a shame that this data isn't available earlier, with the person making the nomination anonymised. Looking back at the historical nominations gives a good feel for the way a nominee's work became recognised over time.

Take, for example, Albert Einstein, who won the 1921 prize in 1922, primarily for work he had done during 1905. His first nomination, a single one, came in 1910. He had two in 1912, and three in 1913. It's possible not only to see the number of nominations grow, but also the prestige of the nominators. By 1920 he had six nominators, including the leading Dutch physicist Heike Kamerlingh Onnes, and in 1921 there were fourteen nominations including names such as Eddington and Planck. Even this didn't persuade the committee, which unusually decided that no one deserved the 1921 prize. But in 1922, the pressure was so great, with seventeen nominations for Einstein, that the Nobel committee gave in and retrospectively awarded the previous year's prize to him.

It's likely that there had been a similar building pressure to recognise the discovery of graphene over several years leading up to 2010 – but we won't know for certain until 2060. We do know, though, that nomination forms would have been sent out around September 2009 to approximately 3,000 professors located all over the world and that from these forms, around 300 people may have been nominated by the February 2010 deadline. After consultation on a number of potential candidates, the Nobel committee would have put forward a report with a shortlist during the summer, and the winners would have been selected from the final candidates by majority vote in October 2010.

It's typical of Andre Geim that, when asked about the benefits of winning the Nobel Prize, he didn't mention the considerable cash sum or the academic kudos. Instead he said that when the vice-chancellor at Manchester had asked the new superstar what he would like as a reward, Geim requested a better parking place. 'It was a fifteen-minute round trip to my car. But it changed that afternoon!'

It's not often that something with an immediate practical application wins a Nobel Prize in Physics. Perhaps the nearest equivalent to the graphene award was the 1956 prize, won by William Shockley, John Bardeen and Walter Brattain for their researches on semiconductors and 'discovery of the transistor effect' – what the rest of us would call the invention of the transistor. In the case of graphene there was certainly a practical innovation, but there were also remarkable new physical properties, arising from its thinness. Because it's thin on a scale that's hard to get your head around.

Plenty of room at the bottom

Before exploring the specifics that make graphene so special, it's worth taking a brief detour to a talk that the great American quantum physicist (and Nobel Prize winner) Richard Feynman made way back in 1959. Feynman's talk 'There's plenty of room at the bottom' was subtitled 'an invitation to enter a new field of physics.' Let's take a look at the opening of Feynman's talk:

> I imagine experimental physicists must often look with envy at men like Kamerlingh Onnes,* who discovered a field like low temperature, which seems to be bottomless and in which one can go down and down. Such a man is then a leader and has some temporary monopoly in

---

* Dutch physicist Heike Kamerlingh Onnes again, who discovered some of the remarkable physics of ultra-low temperatures, such as superconductivity.

a scientific adventure. Percy Bridgman,* in designing a way to obtain higher pressures, opened up another new field and was able to move into it and to lead us all along. The development of ever higher vacuum was a continuing development of the same kind.

I would like to describe a field, in which little has been done, but in which an enormous amount can be done in principle. This field is not quite the same as the others in that it will not tell us much of fundamental physics (in the sense of, 'What are the strange particles?') but it is more like solid-state physics in the sense that it might tell us much of great interest about the strange phenomena that occur in complex situations. Furthermore, a point that is most important is that it would have an enormous number of technical applications.

What I want to talk about is the problem of manipulating and controlling things on a small scale.

Feynman was primarily talking about making physical objects on a small scale, suggesting eventually we could be looking at controlling individual atoms using an army of manipulators. However, what he was describing in that opening section so perfectly describes why Geim and Novolselov's apparently small breakthrough has such significance – they opened up an entirely new field, and though it was taking a different approach to smallness than that envisaged by Feynman,

---

* Far less well known than Kamerlingh Onnes, American physicist Percy Bridgman's work on high pressures led to many discoveries about physical materials, though these arguably have fewer practical applications than the extremes of low temperature.

two-dimensional materials such as graphene are very much exploiting the potential of controlling things on a small scale. Because the definitive property of graphene is how incredibly thin it is.

It's thin

When Geim and Novoselov first peeled away a sheet of graphene on their Scotch tape,* one obvious character-istic was that it was thin. Really, really thin. So thin that it couldn't be seen sideways on and was transparent from above. Specifically, and crucially, the graphene was a slice of matter exactly one atom thick, making it the thinnest known material in the universe. This skimpy substance was a layer of carbon in a hexagonal lattice that was no deeper than the size of the carbon atoms themselves.

How deep is that? A sheet of graphene is around 0.3 nanometres from top to bottom, where a nanometre is a billionth of a metre. To put that into context, the thick-ness of graphene is about 60 times smaller than the tiniest of viruses, 3,000 times less porky than a typical bacterium and 300,000 thinner than a typical sheet of paper. That is thin. Although graphene is very strong, such a thin material generally needs supporting – the material's strength comes in when it is pulled in the same plane as the sheet, but it

---

* Scientists are rarely happy for long with a simple-sounding name. The method of using sticky tape to remove a graphene layer is now referred to in academic circles as the 'micromechanical cleavage technique'.

is very floppy because of its lack of structure out of the two-dimensional plane. To date, samples of graphene have been produced ranging from micrometres (thousandths of a millimetre) to nearly a metre in length.

You're never alone with a substrate

To cope with this floppiness, graphene is always used on a 'substrate' – a piece of material that supports it but doesn't interfere with its properties. The substrates for most of graphene's uses tend to be rigid solids, though they can also be flexible materials like sheets of plastic polymer. In fact, this ability of graphene to twist and shape can be a positive benefit beyond working on a malleable substrate, because a sheet of graphene moulds itself to a surface, although this typically results in the graphene forming creases and folds (which drop away when it is removed). This moulding to the surface and creasing is due to van der Waals forces (see below).

The substrates Geim and Novoselov used in the first exploration of graphene were oxidised silicon wafers, which are readily available from the first stage of silicon chip device production. The starting point for an integrated circuit is a silicon wafer, sliced off an ingot of silicon produced by melting the element. The whole wafer is not oxidised – that would essentially turn the silicon back to silica (also known as sand). It's just the surface that is oxidised to help the layers to cohere.

In practice, it has been discovered that sitting on a silicon dioxide layer degrades the performance of the graphene, because the silicon dioxide surface is relatively uneven, resulting in graphene that is a little tangled in its structure. This produces patchy distribution of areas with a perfect zero band gap and makes graphene a less effective conductor than it usually is. Substrates with an intermediate single layer of boron nitride (of which more later) on top of the silicon dioxide for the graphene to rest on work significantly better. The reason that the nature of the substrate surface has such an effect is because of the forces between the graphene layer and the substrate.

These are called 'van der Waals forces', named after Dutch scientist Johannes van der Waals, and are attractive or repulsive forces that we don't usually notice on the scale of objects we can see. At the atomic and molecular level, though, the action of these forces can be very significant. Attraction is caused by slight changes in charge distribution in adjacent atoms, where the statistically determined position of the electrons happens to put more charge briefly on one side of an atom than another, and also due to other quantum effects.

Although the van der Waals forces are tiny for any particular atom or molecule, they can add up to a powerful effect. Many of the gecko families of lizards are able to run up a vertical wall or even a sheet of glass, because their feet have vast numbers of tiny hair-like structures, each of which generates a small van der Waals attraction to the surface. One of Geim's previous side projects was to produce a prototype of 'Gecko tape', an adhesive tape where the stickiness comes not from a glue but from van der Waals forces generated by a gecko-foot-like array of structures on the tape's surface. Prototypes of the tape have proved surprisingly effective, bearing in mind there is no adhesive involved. As far as graphene goes, the van der Waals attraction is nowhere near as strong as that produced by Gecko tape, but it's enough to make the graphene shape itself to an uneven surface.

In practice, the technique of using Scotch tape to remove a layer from a block of graphite proved an effective combination with the van der Waals force between graphene and a substrate. As Geim and Novoselov discovered when they raided their colleagues' bins, the tape usually picks up multiple layers of graphene. But when that tape is subsequently pressed onto a substrate, it leaves behind flakes of graphite, due to the van der Waals attraction. These are even thinner than the layer that was initially removed, as some of the graphene sheets will remain on the tape. If necessary, the process can be repeated, but even on a first application some of the flakes on the substrate will be single layers of graphene.

It might seem an impossible task to separate out which of the flakes are truly two-dimensional, as even several layers are still transparent, but it turned out that when using an oxidised silicon substrate, there was a clear visual difference between single layered graphene and multiple layers. The way the light reflects back through the layers produces different coloured effects depending on how many layers are present, making it possible to isolate the pure flakes of two-dimensional graphene.

As we've seen (image on page 16), if you could zoom in with a magical ultra-microscope to see the atoms and bonds in graphene, it would look a little like chicken wire with a repeating hexagonal pattern. At each of the six corners of each hexagon sits a carbon atom. Like the material itself, these hexagons are small. Each side is around 142 picometres in length. A picometre is a thousandth of a nanometre, so the sides of the hexagons are 0.142 nanometres long – around half the thickness of the sheet.

Earlier we discussed the 'two-dimensional' claim made of graphene, and concluded that although it may not be two-dimensional in a pure mathematical sense, graphene really does share some properties with theoretical two-dimensional objects. It's useful to think of this two-dimensional environment from a simple visual viewpoint, imagining we have a giant sheet of graphene in space that we can fly around and observe. Its two-dimensional nature means, for example, that we will never see the bonds between the atoms crossing each other. And that means you can't have a knot-like structure, with one bit of the 'string' passing over another, as you can't make a knot in two dimensions. Similarly, two-dimensional graphene can never form structures that require any venture into the third dimension, limiting the way that particles within the space formed by the two-dimensional object can interact and go past each other. What may at first seem like a trivial difference in conformation would result in remarkable capabilities being discovered as graphene was further examined, as we'll see.

The Eureka moment

Just producing graphene was a remarkable breakthrough in

itself, but what would change the Manchester discovery from something merely interesting into a whole new potential field of applications came when the team began to try out the substance's physical capabilities. Geim describes his Eureka moment with graphene coming when they had got to work on its electrical properties. This was easier said than done. It's one thing to have an ultrathin slice of graphite on a piece of sticky tape or oxidised silicon substrate; quite another to be able to test its reaction to electricity.

Novoselov and Geim used tweezers to transfer one of their thinnest flakes to a pristine substrate, then applied tiny spots of silver paint to make electrical contacts to the material. This was an impressive bit of micro-manipulation. The graphene crystal was only about the width of a human hair, making it around 20 nanometres across, and was not as yet a distinct single layer, with several sheets of graphene still together. Lacking any more appropriate technology, the pair applied the silver paint using a toothpick and steady hands. It took a good number of attempts to get it right. But when they did, the result was impressive.

Not only did the material prove to be highly conductive, graphene's resistance – the ease with which an electrical current flows through it – was modified when it was brought near to an electric field. We've already come across the concept of fields a number of times – for example, when Geim levitated a frog using a magnetic field – but it would be useful to clarify exactly what they are before going any further, as electric fields become extremely important when dealing with small-scale electronics.

Working in the field

The concept dates back to Michael Faraday. Before Faraday became involved in electromagnetism in the 1820s, while it was known that electrically charged materials could attract each other, it was described scientifically in the same kind of way that gravity was thought of – as an attraction at a distance. The mathematical approaches taken by Newton to calculate the force of gravity worked in a similar fashion in dealing with electrical attraction. However, Faraday was no mathematician, but rather a superb intuitive scientist.

In thinking about electricity and magnetism, Faraday imagined lines of influence, stretching out from the electrical charge or from a magnetic pole. As an impressive experimenter, Faraday was well aware of the way that iron filings lined up to run from pole to pole of a magnet like a series of contours. These lines, Faraday thought, were a measure of the strength of the force field that the magnet produced.

This idea held that a 'field' was something that spread throughout space and had a value at every location. When something else moved through the field, if it was the right kind of material, it would interact with the field. So, for example, moving a conducting wire through a magnetic field, the wire would repeatedly cut the lines of force in the field and this cutting, Faraday suggested, produced the electrical current that would flow through the wire.

When the Scottish physicist James Clerk Maxwell formalised the mathematics of electromagnetism some four decades after Faraday's first work on the subject, he took

Faraday's descriptive science and turned it into a clean, clear mathematical representation of the way that electric and magnetic fields behaved, how they interacted and how a particular type of interaction between them could produce a wave that travelled at the speed of light. Once Maxwell had done this it was no longer necessary to consider mysterious forces at a distance to deal mathematically with the forces of electricity and magnetism. Maxwell's formulation made fields central to the consideration of electromagnetism, as they remain to this day.

One essential aspect of the electric field is that it will influence the flow of electricity in a nearby body – a process known as a field effect. This is the basis of many of the transistors in chips used in vast numbers of electronic devices. These are usually 'MOSFET' devices: metal oxide semiconductor field effect transistor. The part of the transistor that controls the flow through it, the gate, is used to produce an electric field. It is separated by an insulator from the rest of the transistor, which stops current flowing from it, but the field influences the current flowing between the other two terminals of the transistor, reaching it through the insulator.

Despite being such a crude, hand-constructed device, the graphene crystal that Geim and Novoselov were testing changed electrical resistance by several percentage points when an electrical field was applied from a separate source. As Geim put it: 'If those ugly devices made by hand from relatively big and thick platelets already showed some field effect, what could happen, I thought, if we were to use our thinnest crystallites and apply the full arsenal of microfabrication facilities?'

For months, the Manchester team worked on reducing the thickness of their samples until they were able to repeatedly produce true single layers of graphene. They also expanded the size they could make, from the width of a hair up to around a millimetre across, producing over 50 trial electronic devices. Their paper on producing graphene and a first exploration of its properties would be accepted in September 2004 by the prestigious journal *Science*, after it had first been rejected by the rival *Nature*. The editors at *Nature* felt that there was not enough originality in the work. It's likely that they later regretted this decision.

Not the way it should be

One of the biggest surprises in the development of graphene was that it could be made at all. It had been assumed by physicists for decades that it was physically impossible to make such a thin material.* Attempts to make atom-thin layers of metals by evaporation and deposition resulted in unsatisfactory patchy blobs, rather than a continuous layer. And when attention turned to carbon, theory predicted that for sheets containing up to around 24,000 atoms, the lattice would be unstable and would tend to curl up to form three-dimensional lumps. What's more, as we've already

---

* There's an echo here of the development of the laser at the end of the 1950s, when the big players assumed it was impossible to use a ruby to make a laser due to an incorrect set of data they didn't check properly – but a lone player, Theodore Maiman, used rubies anyway and created the first laser in May 1960.

seen, it was assumed that at room temperature, thermal vibration would rip the graphene apart. In practice, the largest two-dimensional carbon molecule ever synthesised (as opposed to removed from a three-dimensional block as Geim and Novoselov did) was made up of just 222 atoms. Larger sheets were also predicted to be unstable over and above the thermal vibration damage, as it was assumed that inter-atomic forces would result in the sheets curling up to form tubular whiskers.

To make matters worse, the traditional mechanisms for growing crystals would require very high temperatures to make graphene, making the impact of thermal vibrations even greater. And then there was the whole aspect of inter-action with the environment. This is one of the ways that a two-dimensional material is so different from a conventional crystal. If you think of a typical sheet of graphene in a block of graphite, it has other graphene sheets on either side of it, protecting it from reaction. In a sheet of material that is a single atom thick, every atom in the crystal is directly exposed to its surroundings. Any air molecules, moisture, reactive compound and general contamination will hit it full on, and from not one but two directions, as the same individual atoms are exposed on both surfaces of the material.

Yet faced with all these obstacles, and without any precautions to prevent it curling up or disintegrating, sep-aration of graphene flakes using the sticky tape method just worked. It seems that the way graphene is typically pro-duced from a three-dimensional block overcomes some of the issues of attempting to form a two-dimensional layer from scratch. And because its layers are easily removed at room temperature, there is less tendency for them to scroll up, while, once on a substrate, the van der Waals forces tend to keep the graphene pressed flat against the surface and safe from damage. There is some weak reaction with the air – but not enough to destroy graphene's remarkable capabilities.

Beyond the tape

Although 'exfoliation' methods like the original Scotch tape production approach are still used to get hold of small flakes of very pure, single-layer graphene, it isn't an ideal mechanism for larger-scale requirements. It's hard to imagine mass production of graphene-based devices if it all had to start with pieces of sticky tape being repeatedly applied to a graphite block and a substrate.* A number of alternative production methods are now in use which, though often tending to produce less consistently perfect samples than the tape and block method, make it possible to manufacture much larger continuous sheets of graphene and, in principle, should be able to produce it to any size required.

Perhaps the simplest method, which is still hand-crafted but can make a larger sample, is effectively to stitch smaller graphene flakes together. A number of flakes that have been produced via the traditional method are first oxidised, which makes it possible to suspend them in water; the water is then passed through a filtering membrane, which has holes in it

---

* Though there is something rather appealing about the thought of a whole factory filled with row after row of robots, each applying pieces of sticky tape to graphite and then to a silicon oxide wafer.

that are large enough to allow water through but catch some of the graphene flakes. A number of flakes get caught on the membrane, which gradually builds up a graphene layer that can be moved onto the usual silicon or equivalent substrate for support.

Apart from being relatively low-tech, the other advantage of this approach is that the graphene produced is of relatively high quality as it is still the lifted layers from graphite. However, it is very difficult to get a uniform single layer this way, while the graphene oxide needs to be treated to get it back to graphene, a process which itself can introduce irregularities into the graphene. And there is still the original sticky tape step in the process, so while it can be useful to make larger samples in the laboratory, with relatively low expenditure on kit, it is unlikely to became a large-scale production method.

An alternative approach is to produce so-called epitaxial* graphene. This is done by heating a block of silicon carbide to around 1,500°C. This material, also known by the rather magnificent name carborundum, has been produced for over 100 years. It was first used as an abrasive and in cutting discs, but now turns up in more high-tech applications such as ceramic brakes on high-performance cars, in LED manufacture and in steel production. Because of the large market, high-quality silicon carbide is relatively low-cost. When the surface reaches a high temperature the exposed silicon atoms

---

* 'Epitaxial' refers to materials grown by epitaxy, which is growing a crystal on a substrate that determines the orientation of the new crystal. The structure of the substrate acts as a kind of template for the crystalline structure.

boil off, leaving a single atom carbon layer, which can be stripped away as graphene.

However, a more controlled approach that can produce even larger uniform graphene sheets is to heat up a carbon filament in a very high vacuum. Just as the old incandescent light bulbs with a heated filament would leave a deposit of the material that the filament was made from on the inside of the glass, so the glowing carbon filament sprays carbon atoms into the vacuum to land on a metal substrate and form a layer of graphene. This approach can produce large, high-quality films, but it does require room-sized and expensive equipment to produce an extremely high vacuum encompassing a large enough space to accommodate the sheet of graphene.

A more recent variant, called chemical vapour deposition, involves heating a sheet of copper to around 1,000°C at a low pressure, although not requiring such a high level of vacuum as the carbon filament method. A combination of methane and hydrogen is then passed over the surface of the hot copper. Catalysed by the hydrogen, the methane and the copper react, leaving a layer of carbon on the surface. If the sheet is then quickly cooled, the carbon crystallises into a sheet of graphene.

Because chemical vapour deposition does not need an extreme level of vacuum, this is a cheaper mechanism than the approach with a carbon filament, but the quality of the graphene tends not to be as good, as the film of carbon is more likely to pick up impurities from the gases that pass over it. However, it does seem possible to keep these impurities down enough to make a graphene layer of much greater size, coming close to the quality of the flakes from the Scotch tape method. The other challenge with this method is how to get a smooth sheet, as the graphene tends to wrinkle during the cooling process. This is because the copper shrinks at a different rate to the carbon as it drops in temperature. There is still work to be done on improving this method, but it holds out promise for a graphene mass-production mechanism.

Although none of the approaches is yet perfect, there are now several ways to make graphene sheets large enough to be able to produce solar cells or complex electronic devices at a significantly lower cost than, say, silicon equivalents. Large-scale production tends still to be relatively low-quality – but there is no more of a problem here than faced the silicon industry when producing high-quality silicon wafers. That took several decades to perfect – the chances are, with both the wider range of applications and the experience with silicon, that the same quality improvements could be achieved much faster for graphene.

The scene is set for the potential applications of graphene to take off, many of which revolve around its remarkable electronic properties.

Dancing the light electric

As mentioned previously, one of graphene's most remarkable abilities is its extremely high conductivity. This occurs because of a peculiar effect of the graphene sheet's crystal lattice. To get a feeling for this, we need to take a further plunge into the area of physics called band theory. This refers to the workings of the band structure of a substance. As we have seen (see page 56), when a solid conducts electricity, it depends on electrons being able to pull free from the atoms in the substance to conduct the electric current. And the band structure defines how well a particular substance will be able to allow those electrons to act freely.

When atoms come together to form a structure, such as the carbon atoms in graphene's hexagonal grid, the atoms get close enough to each other for their orbitals to overlap and interact. As we have seen, the particular crystal structure of graphene results in an unusual band structure where there are places where the conduction and valence bands exactly touch. This results in the ability of the electrons to interact with vibrations in the lattice in a way that enables the combination of the two to produce charge-carrying 'quasi-particles' which effectively have no mass. Usually, when electrons are the charge-carrier, they travel quite slowly through a conductor, floating along at a walking pace. (The reason electricity doesn't take a long time to get from one end of the wire to the other is that an electromagnetic wave travels down the conductor at the speed of light, setting all the electrons in motion as it reaches them.)

The slow pace of electrons in a normal conductor is due to their frequent interactions with the electrical charges in the atoms of the material. But in graphene, the charge-carriers, acting almost as if they were photons of light, can get up to speeds of around a million metres per second (for comparison, light in a vacuum travels at around 300 times this speed). This seems almost impossible, but the combination of the electrons and the crystal lattice acts as if the charge is being carried by massless particles that are travelling through it at high speed.*

Where electrons usually find that the atoms in a lattice act like a set of barriers that slows them down, the quasi-particle tunnels through the barriers as if they aren't there – this is quantum tunnelling, as described in Chapter 3. It's a bit like the difference between two people trying to move forward in a set of jumps, one on a concrete floor, the other on a very long trampoline. It's the interaction between the gymnast's muscles and the spring structure of the trampo-line that generates what would otherwise be impossibly long jumps and fast speed – similarly, the interaction between the electrons and the two-dimensional structure of graphene generates the otherwise impossibly fast charge-carriers.

Because the charge-carriers are moving so quickly, phys-icists have to switch the equation used to describe their behaviour. As we have seen, when anything is moving as quickly as these particles, the impact of special relativity has to be brought in, which is why in the previous chapter we were introduced to the Dirac equation – this rules the roost in graphene. And it is these relativistic charge-carriers that result in graphene being a far better conductor than copper, silver or gold.

---

* The massless quasiparticles are formally known as massless Dirac fermions and have the rare honour of appearing in a TV comedy pro-gramme – specifically the episode of *The Big Bang Theory* titled 'The Einstein Approximation', in which the character Sheldon is seen puzzling over their behaviour.

However, with graphene, the surprises keep coming. It has recently been discovered that this isn't the end of the story. Electrons themselves in the graphene behave in a way that is quite unlike the way they travel in a metal. Individual electrons interact with each other. This happens normally, but the usual result is bouncing off each other and randomly scattering, reducing the ability to carry current. In graphene, the electrons form a kind of gooey electron fluid, which has a viscosity (resistance to movement) that is as much as 100 times that of honey at room temperature. This remarkable and previously unknown behaviour means that the electrons can form whirlpools and eddies inside the graphite like water in a river. Sometimes they have even been observed moving in the opposite direction to the electrical current.

Such perverse behaviour of going against the flow in a systematic way has never been seen in electrons before. This so-called 'electron hydrodynamics' is fascinating in its own right and is proving a new avenue of exploration for theoretical physicists who hope to understand better the mechanisms of electrical conduction in solids. However, it also has a surprising side-effect, which is that it can make the graphene even better as a conductor than it would be without it.

This is counter-intuitive, as you would expect that electrons moving backwards against the flow of electrical current would reduce the material's ability to conduct. What seems to be happening is that the electrons that form the slow-moving, eddying fluid flow stay near the edges of the material. These rivers of electrons provide a pair of repulsive barriers that prevent the faster moving electron/lattice charge-carriers from being held up by collisions with the other electrons, making it possible for the graphene to exceed the theoretical limit for the amount of current that it could carry.

This property was only reported in late 2017 – graphene continues to amaze long past its original discovery. Just as Richard Feynman hoped, opening up the world of the very small has not just made a new scientific discovery possible, but makes so many new areas of exploration available.

In the Hall of the quantum king

There is one extra peculiarity about the electronic behaviour of graphene, for which we need to introduce a little more quantum physics to get a picture of what's going on. The effect in question has the impressive name 'the quantum Hall effect', while also of relevance is 'the anomalous quantum Hall effect' – the latter displayed not by graphene but by other ultrathin materials. Let's deconstruct the names. The Hall effect, which predates quantum physics, was discovered by US physicist Edwin Hall in 1879.

If you put an electrical current through a conductor and add a magnetic field from one side, the flow of electrons through the conductor will not be in a straight line, but will curve as a result of the magnetic field. This means that there will be more negative charge on one side of the conductor than the other, which in turn means that there is an electrical field set up between that relatively negative side and the other side which will be relatively positive.

The next step is the quantum Hall effect, which, as the name suggests, introduces quantum behaviour. This happens in two-dimensional conductors or semiconductors when the temperature is very low (within a few degrees of absolute zero, which is −273.15°C) and the magnetic field is strong. Under these conditions, the resistance of the object at right angles to the electron flow becomes quantised – it can only take on very specific values. To be precise, the only options are limited by two constants of nature – Planck's constant h, which gives the relationship between a photon's energy and its wavelength and e, the charge on an electron. The values observed are specified by a variable named v, which can take a range of integer or fractional values, giving the object a resistance of h/ve^2.

These narrowly prescribed values of resistance, which are extremely precise, make the quantum Hall effect very useful for devices requiring an exact resistance to make electrical measurements, so it is valuable in various kinds of detector. More interesting still, the resistance of the material in the direction of the current flow disappears. Electrons flow along the edges of the material without any losses to resistance. This means that, in principle, there would be none of the energy loss to heat we get in an ordinary wire conducting electricity.

Interesting though the conventional quantum Hall effect is, it is of little practical value because it's not realistic to have wires that are kept in a powerful magnetic field at ultra-low temperatures – a fraction of a degree away from absolute zero – for everyday applications. It's fine for the lab, but it's not going to be used in a commercial device or in wiring. However, graphene's weird conductivity means that it can produce the quantum Hall effect at room temperature, though it still requires the strong magnetic field.

Going a final step, we get to the anomalous quantum Hall effect – which is the trick performed by some other ultrathin materials, but not by graphene. The particular type of ultrathin substance is called a magnetic topological insulator, which acts as an insulator in its interior, but conducts on the surface. Bearing in mind the quantum Hall effect influences the edges of the thin material, these seem an ideal constituent with which to use the quantum Hall effect – and they are so good at it that the result is the so-called quantum anomalous Hall effect where the quantum effect occurs without a magnetic field.

In tests with an ultrathin film of a material made from bismuth, antimony and tellurium, doped with chromium, experimenters at Stanford and MIT in the US and the Tsinghua University in China have produced a near-perfect anomalous quantum Hall effect across the material with a low resistance of about 1 ohm lengthways. As yet this only works at ultra-low temperatures. However, the two different approaches with graphene and these special compounds individually overcome one of the limitations of the quantum Hall effect each – in the future, it's entirely possible that some combination of ultrathin materials could make it useable without a magnetic field and at room temperature.

Superstrength

One of the most remarkable claims for graphene is that it is far stronger than steel. In fact, it's currently the strongest substance that has ever been tested. We need to give a little clarity to that remark – the invisibly thin sheets of graphene can't compare with the strength of a centimetre-thick sheet of steel. You can't lift an elephant using a single two-dimensional sheet of graphene. The problem is that the term 'strength' is rather loose in this context.

The claim that graphene is the strongest material ever tested refers to tensile strength, which is the material's ability to resist being pulled apart lengthways.* And the standard measure of tensile strength (in which graphene stands out way above its possible competitors) requires an equal cross-section of material for like-for-like comparison. This means that to make a fair comparison we need to put layers of graphene into a composite with some kind of binding material to make it up to the required thickness.

Tensile strength is measured in pascals (equivalent to newtons per square metre), which is more commonly the unit of pressure. (Because the numbers are large, it's more common to measure it in megapascals, where 1 megapascal = 1 million pascals.) To get a feel for scale, the typical tyre pressure of a car is around 0.2 megapascals. The table opposite shows how graphene stands up to the competition.

---

* Although many requirements for strength aren't just about pulling the substance, tensile strength often reflects ability to stand up to other stresses and strains. So, for example, a bulletproof vest's ability to stand impact depends on how easy it is to stretch the material in the vest at right angles to the bullet's direct of travel, as the bullet will try to stretch the material apart to get through it.

Substance	Tensile strength (megapascals)
Graphene	130,000
Boron nitride nanotubes	33,000
Silicon (monocrystalline)	7,000
Limpet*	5,000
Kevlar	4,000
Diamond	2,800
Strong steel	2,500
Brass	500
Human hair	225
Pine	40
Iron	3

Various tube-shaped variants of graphene (carbon nanotubes) also have extremely high tensile strength, though in the context of putting large numbers of them in a composite, these are effectively just alternative ways of structuring the graphene.

The reason that graphene is ridiculously strong is primarily down to its bonds. Looking at the lattice structure of graphene (see page 16), it has vast numbers of carbon–carbon covalent bonds all arranged in the same direction. A square metre of graphene, weighing in at an impressively light 0.77 milligrams, contains around 10^{20} atoms,** each with three bonds attached to each atom.

---

* This is not some strange material, but the amount of pull required to dislodge a limpet when it is stuck to a surface.
** It's hard to visualise 10^{20} – it is 100,000,000,000,000,000,000 atoms.

The covalent bonds between carbon atoms are strong bonds, and combine ideally here to resist being pulled apart, just as in the different crystalline structure of diamond the lattice of covalent carbon–carbon bonds gives the material its hardness. A good piece of graphene is also unusually low in faults in the lattice structure. Anywhere the neat repeating pattern of the bonds is broken is an opportunity for a split to start forming in the material – but compared with a metal like steel, good quality graphene has far fewer such faults.

This remarkable tensile strength makes graphene an ideal choice for the reinforcement in future composites, with the graphene embedded in another material, usually a plastic polymer, to give it extra strength. Like carbon fibre or carbon nanotubes, graphene isn't great at sticking to the composite material, but it could be treated chemically (for example by converting it to fluorographene – see page 111) to make the interface more effective. Graphene is not only stronger than carbon fibres, but because it is a single atom thick, it can't split in the dimension at right angles to the two-dimensional sheet, increasing its effective strength and making it excellent at stopping crack propagation.

Because, as we have already seen, graphene is so conductive, it is only necessary to include a small percentage of graphene in a plastic – around 1 per cent of the whole material – to make the plastic conducting, which then gives it a whole range of extra potential applications. This can be done using cheaply produced flakes of graphene just millionths of a metre across. Similarly, powdered graphene could be used to replace the graphite or carbon fibres in batteries, significantly increasing their efficiency.

A sensitive surface

One surprising possible use for graphene is in providing ultra-sensitive gas detectors that can pick up the presence of a single atom of a gas. Such a detector would involve an exposed graphene surface, onto which gas molecules would adsorb – effectively sticking to the surface. Because graphene is such a superb conductor, and its conduction is due to interaction between its crystal lattice and conduction band electrons, even a single atom sticking to it will make a small change to its conductivity, which can be picked up and analysed.

In tests, graphene detectors were inserted into a glass tube which contained either helium or nitrogen, plus a range of contaminants. In the first trials, for common air pollutants from nitrogen dioxide to carbon monoxide, concentrations of one part per million were easily detected from a change in the current flow, with nitrogen dioxide detected almost immediately the pollutant was added. By the end of the tests it was possible to detect the impact of individual gas molecules, making detection in fractions of parts per billion possible.

This kind of application is just the start, though, of graphene's capabilities, particularly once the existence of other two-dimensional materials is added into the mix.

OTHER FLATTIES

5

There's more to two dimensions than carbon

Graphene sprung to the fore as a result of Geim and Novoselov's work, but once they had proved that it was possible to make stable two-dimensional materials, graphene was never going to be the only kid on the block. Carbon is not the only atom capable of forming two-dimensional sheets, particularly once compounds are considered. From boron nitride to molybdenum disulfide and the mysterious sounding dichalcogenides, the field of the ultrathin is proving unstoppable.

Graphene goes white

Perhaps the best-known of graphene's rivals* is boron nitride, which we've already met as one of the few substances to come

---

* More accurately you could describe boron nitride as a colleague rather than a rival as it is often used alongside graphene.

close to graphene in tensile strength. In its two-dimensional form, boron nitride is sometimes known as 'white graphene'; despite having a totally different chemical structure and very different properties from the carbon-based equivalent, it is another winner of the ultrathin world. Boron nitride is a simple inorganic compound made of pairs of boron and nitrogen atoms. This combination has a similar versatility to carbon in its ability to make bonds. The result is an equivalent set of allotropes with structures similar to diamond and fullerenes. Most importantly here, though, boron nitride can form single-layer hexagonal lattice sheets like graphene, which will initially form a block equivalent to graphite.

In its two-dimensional sheet form, like graphene, the atoms of boron nitride are arranged in a hexagonal lattice, but with the significant difference that atoms around each hexagon alternate between being boron and nitrogen. Each atom is connected to one of its neighbours by a double covalent bond and to its other two neighbours by single covalent bonds. With four bonds per atom, there are no freely available electrons, meaning that hexagonal boron nitride has a wide band gap and is a good insulator.

The hexagonal lattice of boron nitride.

The combination of a similar physical structure but different electron availability means that when boron nitride takes this hexagonal form resembling graphite, it shares some properties with graphite, while being very different in others. For example, like graphite it is a good lubricant. Because it also doesn't react well with other chemicals, it has gained an unlikely role as a lubricating agent in cosmetics, as well as having more traditional applications such as being embedded in ceramics designed to withstand high temperatures and in providing the slipperiness for self-lubricating bearings. It has even joined graphite as a component of some pencil leads – it doesn't make a good writing material alone, but it renders the reconstituted graphite more stable. But of particular interest here, as with graphite, is the weakness of the bonds between layers. It can shed single atom-deep layers: the hexagonal boron nitride equivalent of graphene.

One possible use for nano-scale sheets of boron nitride is to reduce water pollution. A sheet made up from the boron nitride layers can absorb up to 33 times its own weight in potential pollutants such as oil and organic solvents. However, the boron nitride repels water molecules, so it proves very effective at removing these kinds of pollutants from water. Once the sheet has become saturated, it can be cleaned by heating up to a high temperature, staying in one piece as the pollutants burn off. The ability to repel water also means that a layer of boron nitride could be used in self-cleaning electronic display screens, which won't get misty with condensation when the air is humid.

As with all the two-dimensional compounds, one of the most far-reaching potential applications of boron nitride sheets is in electronics. Combining layers of the insulating boron nitride with layers of superbly conducting graphene is a recipe for all kinds of electronic devices. Because we are operating at the scale of atoms, quantum effects can be very strong between the different layers. This means, for instance, that with single boron nitride sheets quantum tunnelling (see page 50) can take place, giving the potential to make tiny components making use of this effect. With a number of boron nitride sheets between a pair of graphene sheets it becomes possible to construct miniature high-capacity electrical storage devices, as we will discover in the next chapter.

As with all these kinds of multi-layer concepts, we are currently in the early days of development, but the combinations are exciting indeed to those whose job it is to produce smaller and smaller electronic components.

Slippery moly

The compound molybdenum disulfide is probably most familiar to engineers as 'Moly' (pronounced molly rather than moley), an additive for grease lubricants that reduces wear on the equipment it lubricates. This naturally occurring compound isn't quite as thin as graphene in its thinnest hexagonal layer form – rather than a single atom layer, it's three atoms thick, as a central layer of molybdenum atoms link to sulfur atoms that stick out either side of the sheet. However, it's still thin enough to count practically speaking as a two-dimensional substance.

The ultrathin version of molybdenum disulfide was not discovered until 2011, produced by the same Scotch tape peeling technique as graphene, though it is now also produced chemically, grown on a silicon wafer. Like the thin layers of carbon in graphene, the molybdenum disulfide layers move easily over each other, giving it its lubricating properties.

Also like graphene and boron nitride, molybdenum disulfide is interesting electronically, but in this case, it completes the set of useful ultrathin materials by being a semiconductor, sitting between graphene's impressive conductivity and boron nitride's insulating capabilities. The band gap of molybdenum disulfide is valuable as it's just right for the change in energy between the conduction and valence bands to match the energy in a light photon, so molybdenum disulfide has plenty of potential both as a light detector, where a photon is absorbed as an electron jumps up in energy from valence to conduction band, and as a light source when an electron jumps down into the valence band.

Perhaps the biggest potential, though, is as an alternative to silicon in developing a much wider range of solid state electronics. Production of traditional silicon-based electronics is getting ever closer to its physical limits of miniaturisation, but in October 2016 what is currently the smallest ever transistor was made with molybdenum disulfide and carbon nanotubes. The transistor's active part, the gate, was just 1 nanometre across – contrasting with the 20 nanometre gates in the smallest commercially available silicon chip transistors.

By December 2016, researchers from Stanford had moved on from single transistors to show how sheets of molybdenum disulfide can be used to make practical electronic circuits on a scale smaller than their silicon equivalents. It's still early days for this material, but it is rapidly moving from an experimental material to a mass-market contender.

As with the other two-dimensional films, the physical properties of the transparent sheets of molybdenum disulfide make for extremely thin transistors, which could be built into any sheet of glass. This would make it possible to turn windows, car windscreens or spectacles into information displays. And, once again, the sheet is flexible, which makes molybdenum disulfide one of the options for circuitry that can be built into paper-like screens, go-anywhere solar panels, clothing and more. And the structures aren't limited to simple sheets – like carbon, molybdenum disulfide can form fullerene structures, including nanotubes. These are showing promise as electrodes in experimental high-performance lithium ion batteries.

In parallel, like its fellow thinnies, the structure of molybdenum disulfide is proving interesting as a filter – in this case, for producing drinkable water from seawater. For some time, experiments have been undertaken using graphene as a porous membrane to allow water through but to block the flow of salt ions in a mechanism called reverse osmosis. After a number of thin film materials were computer modelled for this role in 2015, molybdenum disulfide came out ahead, coping with more than half as much water again as a graphene filter. This benefit seems to be because the pores allowing water through tend to be surrounded by molybdenum, which pulls the water towards the pore, while the adjacent sulfur atoms push the water away, encouraging it to clear the pore and move beyond.

These three leaders of the early ultrathin revolution – graphene, boron nitride and molybdenum disulfide – are likely to provide the backbone of the first generation of ultrathin applications, but they are by no means the only contenders.

From silicene to dichalcogenides

Study the papers being written on the ultrathin and you will find a number of other names cropping up regularly. One is silicene. The element silicon is probably the closest in behaviour to carbon – some scientists have even suggested that there could be silicon-based life somewhere in the universe to complement our familiar carbon-based lifeforms. So, it seems reasonable that silicon could form a structure something like graphene, and it does – given the name of silicene.

Unlike graphene, silicene does not occur naturally and was not discovered until 2010, when the substance was produced in small quantities by depositing silicon onto a silver substrate. And there are significant differences in the way the structure forms. While silicene has the familiar hexagonal lattice, it isn't absolutely flat like carbon's, giving a buckled structure that makes it, for instance, less useful as a lubricant, but that can have actual benefits for electronics applications.

The crumpled surface results in a band gap that can be modified easily with an external electrical field, and its atomic structure makes it much easier to react with doping agents. This means that silicene may be better than graphene for producing field effect transistors in an integrated circuit – so graphene isn't necessarily the end of the road for the silicon chip. One essential that hasn't been fully explored is what will be the best substrate for silicene to operate on – as yet, these have been expensive materials compared with the substrates used with the core ultrathin materials.

Other ultrathin possibilities are represented by the mysterious-sounding compounds dichalcogenides. In practice, these aren't as exotic as one might think. A chalcogen is a fancy name for an element in group 16 of the periodic table, which includes oxygen, sulfur, selenium, tellurium and polonium. A chalcogenide is a compound of one of these with another element, typically a metal (by convention, oxygen tends not to be considered as forming a chalcogenide).

The most likely forms to be used are transition metal dichalcogenides, where the other element is from a particular block of the periodic table, such as molybdenum or tungsten. Molybdenum disulfide, described in the previous section, is in fact a transition metal dichalcogenide, and a number of other compounds in this category are also showing promise for electronics, with the right-sized band gap for emission of photons as a light source or absorption of photons as a detector. These include tungsten disulfide, molybdenum diselenide and molybdenum ditelluride. Similarly, this kind of application can be used in field effect transistors (see page 65) – increasing still further the range of ultrathin materials available to produce extremely compact and flexible electronics.

Some of these materials, for example tungsten ditelluride, have properties that make them suitable materials for experimental 'spintronic' devices. Normal electronics deals only with one property of the electron – its electric charge. However, the electron has other properties too, notably its spin (see page 59), which is a quantum property with a value of either up or down when measured in any particular direction. Although it's early days, a lot of effort is being put into spintronics as it would make it possible to pack more information into a single bit, without all the complexity required to make quantum computing* work.

As semiconductors, the transition metal dichalcogenides are very much complementary to graphene – but the final option to go further with ultrathin materials returns to graphene itself, with an added twist.

Compound interest

In effect, a sheet of graphene is a single, enormous carbon molecule, which can undergo chemical reactions to produce a new two-dimensional substance like any other molecule. To date this has been done with hydrogen to produce graphane, with one hydrogen atom per carbon atom, and with fluorine to produce fluorographene, with a fluorine atom per carbon

---

* In quantum computing, rather than bits, the computer works with qubits, where each qubit is the combined states of a quantum particle. The capabilities of the computer increase exponentially as you add qubits, but as yet it has proved difficult to make a stable, large-scale quantum computer.

atom. Both compounds are stable substances, though fluoro-graphene seems the more robust of the two and has been explored more to date.

Fluorographene retains the familiar hexagonal lattice of graphene, but with a fluorine atom attached to each carbon. As yet, the best way to produce this seems to be exposing a sheet of graphene to XeF_2, one of the rare compounds of the noble gas xenon. Fluorographene has only been made in small quantities so far, but the evidence is that it is an excellent insulator – so may be useful in multi-layered con-struction with highly conducting conventional graphene. The degree to which a compound is an insulator depends on the size of the band gap (see page 56). Graphene has no band gap, while fluorographene has a very wide one. It is specu-lated that other graphene compounds could sit between the two, giving them properties more like a semiconductor. This would provide a wider range of substances that could be used in ultrathin electronics.

Fluorographene is also able to resist temperatures up to around 200°C and is chemically inert. Apart from acting as a two-dimensional insulator, fluorographene could provide a thin film equivalent of polytetrafluoroethylene (PTFE), which is a long chain carbon molecule with fluorine atoms attached. It was originally made by accident in 1938 when an American chemist, Roy Plunkett, was trying out different compounds as possible new refrigerant gases. He was using a cylinder of tetrafluoroethylene gas – a simple compound comprising a pair of carbon atoms and four fluorine atoms. The gas in the cylinder seemed to have run out, yet it felt far too heavy to have nothing remaining in it.

Concerned that a chemical reaction had taken place that could have rendered the gas explosive, Plunkett took the cylinder outside his lab and rigged up a blast shield before cutting through the casing. Inside was a slippery, white, waxy solid – PTFE. The high pressure the gas was at, combined with the iron interior of the cylinder acting as a catalyst, had enabled the polymer to form. PTFE was first used, as it still is, to coat joints to make a good seal, but we also come across it in non-stick pans, where it's mostly known by the brand name Teflon.

Those non-stick pans were first made in France in 1954, when Colette Grégoire suggested to her husband Marc that he find a way to prevent food sticking to pans using the slippery PTFE he had put on his fishing tackle to make it run more smoothly. It proved a non-trivial task, as the PTFE didn't easily stick to the metal, but Grégoire achieved it by first pitting the aluminium surface with acid, then heating PTFE powder on the surface so it gripped the uneven metal. The Grégoires would start a company, Tefal, selling pans with Teflon coating. PTFE naturally repels water and has very little for other molecules to cling on to – it even resists the van der Waals forces (see page 80) that enable a gecko to walk up a vertical sheet of glass.

Conventional PTFE molecules are effectively one-dimensional structures in the form of long chains of linked carbon atoms, each attached to a pair of fluorine atoms. Fluorographene takes PTFE up to two dimensions (admittedly with only one fluorine atom per carbon atom), which may prove particularly beneficial in obtaining extremely even layers for high-tech equipment, or duplicating PTFE's low adhesion abilities on much smaller objects.

The versatile film

Whether using graphene, which continues to have the most extraordinary capabilities of any substance known, or the various alternative two-dimensional materials, there has been an explosion of potential applications of these versatile substances. Many are still at the development phase – not surprising, given that graphene was only first produced in 2004. However, the range is remarkable.

THE ULTRATHIN WORLD 6

With hundreds of scientists working on ultrathin materials around the world, exploring both the underlying physics and chemistry of their two-dimensional structures as well as developing ways to use them, we are already seeing some remarkable products at the prototype stage, even though at the time of writing there is little yet in full production.

Almost inevitably, one of the first areas to be given consideration is the future of ultrathin electronics.

Graphene transistors

Given the observation of the field effect that Geim and Novoselov made with devices constructed from sticky tape, silver paint and toothpicks, it is no great surprise that one application that sprang to mind early was finding a way to produce graphene transistors. With its incredible thinness, a layer of graphene would make for a tiny circuit that could even be flexible if required. Although, of course, it requires a substrate to support the graphene's inherent floppiness, which limits the overall thinness possible, this substrate can itself be thin, and it's possible to imagine thousands of graphene-based circuits, perhaps piled on top of each other with suitable insulating layers.

Field effect transistors (see page 65) made from graphene are of extremely high quality, allowing control over high levels of quasiparticle flows by an imposed electric field, beating the best silicon/metal oxide field effect transistors. They are particularly effective for the high frequency applications which are often required by modern electronics.

There is, however, what appears initially to be a serious problem when it comes to producing integrated circuits based on graphene rather than silicon. To work together to provide the logic gates of a computer chip, for example, the transistors in the tiny circuitry have to be able to turn flow on and off – to act as switches, rather than as amplifiers. As we have seen, graphene is a wonderfully good conductor, so much so that it is very difficult to avoid leakage of quasi-particles even when their transport is suppressed as much as possible.

This is not an impossible problem to overcome, though. One approach that has been tried, with success, is to use graphene-based structures, such as multi-layered ribbons or 'quantum dots' which aren't made up of a pure sheet of graphene, but can exhibit more of the on/off control required for a logic gate and still have many of graphene's benefits.

Another option would be to selectively react carbon atoms with fluorine. As we have seen (page 112), fluorographene, with a fluorine atom attached to each carbon atom, is another potentially useful ultrathin film. And it is possible in principle (though not yet in practice) to set up a hybrid of graphene and fluorographene by adding fluorine to selected carbon atoms in the lattice. Fluorographene is a very effective insulator, and adding its insulating properties to limit conductivity could make it possible to devise a fully operational circuit complete with logic gates all embedded within a single sheet of graphene.

Finally, there is the possibility of using dual-layer graphene. Putting together two layers of graphene does not return it to being boring old graphite, but modifies the properties of the graphene. Depending on the way the two layers line up, it can have interesting electronic properties of its own, which are different to pure graphene. In the so-called Bernal arrangement, where half the atoms in one layer are above other carbon atoms, but the other half are above the centres of the hexagonal spaces, dual-layer graphene or bigraphene has shown promise in having the right kind of band gap to facilitate the production of computer gates. Alternatively, we could still look to lithography, the technique used on conventional silicon chips, but with a new twist, as will be discussed in the next section.

Graphene transistors can not only make use of field effects for amplification, but also provide the possibility of producing ultrafast transistor-based photodetectors, which work across the visible and infrared light spectrum. Not only would such detectors be flexible in the ways they can be used, they would literally have the potential to be flexible, providing light detection facilities and potentially the mechanism for a miniature camera that can be adapted to any shape and size.

Working two-dimensional chips

Although there are undoubted problems with using graphene as a basis for integrated circuits, there are obvious benefits in using a material that is cheaper, more resilient and far thinner than silicon. By 2011, IBM had announced the first prototype chip that was based on graphene.

The demonstration was a radio frequency mixer chip. These devices take two different signals and produce an output based on some combination of the two – most commonly, a mixer outputs the sum of the two signals. The circuit was made using two to three layers of graphene. To form the circuit, the graphene was first spin-coated with a polymer. This is a process where a dot of the polymer is placed in the middle of a piece of graphene, which is then rotated at high speed – typically around 50 revolutions per second – so that the coating spreads out to form a thin layer over the surface.

The polymer was itself then given a coat of hydrogen silsesquioxane. This may sound like a toothpaste additive, but is in reality a compound of hydrogen, silicon and oxygen which forms a polymer that works well as a resisting material when producing the pattern that is required to form an integrated circuit chip. The structure of the circuit was then produced using electron beam lithography. This is a process using a tightly focused beam of electrons, which changes the structure of the hydrogen silsesquioxane, making it soluble.

The treated parts are then removed, leaving behind channels which will act as a mask so the unwanted graphene can be removed with a laser.

After the lithography was complete, the graphene chip was cleaned using acetone, resulting in a final device less than 1 square millimetre in size. Although this approach isn't suitable for mass manufacturing – electron beam lithography is slow and expensive – it did demonstrate that it is possible to make a functioning graphene-based chip. What's more, the device worked with very high radio frequencies, up to 10 gigahertz. For comparison, FM radio is typically around 0.1 gigahertz, digital radio is around 0.22 gigahertz, and mobile phones operate at around 0.85 to 1.9 gigahertz. The ultra-high frequencies that the graphene chip can handle make it ideal for secure, close-range military communications.

Perhaps the biggest difficulty with pure graphene electronics to date has been that it is just too good a conductor. As we have seen, graphene transistors work fine as amplifiers, but without modifying the molecule, it has proved impossible to get them to switch off entirely – which has limited the application of electronics based on graphene alone. However, in late 2017, researchers at Rutgers University in New Jersey announced that they had discovered a way to make graphene pause its conductivity – which has the potential to provide an alternative route to making fully functional graphene transistors a reality. To do this, they made use of the probe of a scanning tunnelling microscope – a device we encountered briefly in Chapter 1, whose calibration method facilitated Geim and Novoselov's early supply of Scotch tape-graphene samples (see page 14).

Scanning tunnelling microscopes are remarkable devices, which bring a tiny, electrically charged point extremely close to the surface of the material to be observed. When the point is almost touching the surface, electrons from the charged point of the device undergo quantum tunnelling across the barrier formed by the gap between the tip and the material beneath. Any tiny changes in separation between this point and the surface have a strong effect on the electrons' ability to tunnel, so as the tip is moved back and forth over the surface it can plot out an object's surface in great detail.

It might seem that the tip needs to be incredibly sharp to work on this scale – and it does – but achieving this is rather easier than might be expected. Ideally the point should have a single atom of the metal it is made from protruding from its end. The sophisticated technology used to hone the metal to this invisibly fine point is a pair of wire cutters. A thin piece of metal wire is simply snipped with the wire cutters – in practice, there is always a single atom sticking out somewhere on the top, and this works fine as the microscope's tip.

By varying the electrical current applied, the tip of a scanning tunnelling microscope can also be used to move individual atoms – famously, IBM scientists spelled out the company's name in 35 xenon atoms using one of these devices back in 1989. The Rutgers researchers used the same effect to create an electrical force field in the graphene beneath it, which stops the charge-carriers in their tracks or forces them to travel along certain paths while the probe is present, like a lens acting on light.

'IBM' in atoms.
(Image originally created by IBM Corporation)

Although a scanning tunnelling microscope is small compared to traditional electron microscopes, it is still far too bulky to be part of an electronic circuit. However, the vast majority of the microscope is not necessary for this use. All that should be required is a set of tiny wires on top of the graphene, which can set up the electrical field to control electron flow and turn the graphene into circuits containing fully functional transistors.

Spin doctors

As we have seen (page 111), as well as better conventional electronics, graphene and its fellow ultrathin materials are also likely to play a part in the still infant field of spintronics, where not only electrical charge but also the spin of the electrons is used in producing logic circuits.

Researchers in Spain, the Netherlands and Germany, as part of the Graphene Flagship initiative (see page 150), made significant steps forward in 2017 on the kind of development needed to produce a practical spintronic device. Just as it stands out in so many other ways, graphene has been discovered to have a uniquely high 'spin lifetime anisotropy' – in effect the ability to keep spins locked in a particular direction for far longer than normal. The graphene-based device acts as a filter which will only transmit certain directions of spin, allowing it to act as an ultra-sensitive detector of spin changes, necessary to make use of spin as the equivalent of the state of a traditional electronic bit in computing.

When graphene is combined with other ultrathin materials, such as molybdenum disulphide and tungsten disulphide, the interaction between the layers can provide a mechanism for controlling the lifetime of the spin directions before the electrons undergo 'spin relaxation', reverting to random directions – the equivalent of going from a 1 to a 0 in an electronic logic circuit. If reliable spintronic logic devices can be constructed – and the Graphene Flagship's collaboration between researchers and development companies seems ideal to fast-track this – they could benefit both conventional devices and the new field of quantum computing.

However, not every electronic or even spintronic application of graphene requires as complex a structure as an integrated circuit. In some cases, it is enough that it is a superb conductor that just happens to be transparent too.

Light fantastic

Anyone involved in the production of solar cells is certainly keeping an eye on the development of graphene. To make an effective solar cell, producing electricity from sunlight, requires the use of a conducting layer of material that is transparent. To date, the need to combine transparency and conductivity has meant using very thin layers of metal, or metal oxides. However, these tend to be less transparent than graphene,* are expensive to make, are more variable in the frequencies of light that they absorb and tend to be more likely to undergo unwanted chemical reactions.

Prototype solar cells using graphene produced by chemical vapour deposition (see page 90) have been made now for some years and it should not be long before graphene is transforming this industry. What's more, combined with the capabilities of a flexible ultrathin light-absorbing semiconductor such as molybdenum disulfide, graphene could help make far better flexible films that are still able to generate electricity from sunlight, wrapped around any shape of surface.

The same requirement for a transparent conductor is also true of LCD screens and touchscreen technology. These require significantly larger sheets than a solar cell, which is typically only a few centimetres across (solar panels are made up of arrays of multiple cells), which in the early days of ultrathin film production was a problem, but the newer production techniques are making it possible to produce two-dimensional materials on the kind of scale required for modern displays. Bear in mind how LCD displays, which were first limited to a few centimetres across, are now

---

* For a two-dimensional substance, graphene actually absorbs quite a lot of the incoming light, only allowing around 97.7 per cent through, but this is still significantly more transparent than the thicker metal layers that are used in these technologies.

routinely produced for TVs with diagonals of 50 to 70 inches (1.2 to 1.8 metres).

We can expect larger scale graphene panels within a few years. But they are already produced on a scale that could make one of our most familiar domestic disasters less of a problem.

Smashing screens

Anyone who has a smartphone knows how easy it is to break the screen. Drop your phone on a hard surface, and the chances are you will end up with a crazed piece of glass and an expensive repair bill. But a team at the University of Sussex have come up with a flexible alternative. What's more, their robust screen has the potential to use less energy and will be more responsive than a traditional glass touch screen. It could also be used in other devices where a flexible touch connection would be useful.

The Sussex screen uses a flexible acrylic plastic, coated with a grid of silver nanowires and pieces of graphene, which are floated on water, picked up with a rubber stamp and pressed onto the silver grid in whatever pattern is required – it's a bit like potato printing with graphene. The benefits are remarkable. The resultant screen is flexible, and thanks to the remarkable conductivity of graphene is about 10,000 times as conductive as it would be with silver alone, making it more sensitive and far less power-consuming.

The amount of metal in the nanowires used is also reduced hugely from a traditional screen, making the graphene and silver approach significantly cheaper than the current most frequently used alternative, indium tin oxide. And the graphene prevents the silver from tarnishing – always a problem when that metal is used in air.

While the flexible screen, like many of graphene's uses, relies on the substance's electronic special nature, other possible applications may depend on its other properties: for example, how it deals with magnetism.

Magnetic trick

Appropriately, given Andre Geim's history of levitating frogs and other diamagnetic objects, it turns out that graphene too is a diamagnetic material. As we have discovered, this is a medium that itself becomes a (relatively) weak magnet in opposition to a magnetic field that it is exposed to. In effect, a diamagnetic object is repelled by a magnetic field. In the case of graphene, the special electronic structure of the lattice makes it a strong enough diamagnetic material to be levitated by conventional neodymium permanent magnets, not requiring the massively strong electromagnets needed to levitate frogs.

Although there isn't an immediate application for this effect – we aren't going to see graphene holding maglev trains above the track as a result of diamagnetism, as the repulsive force is too weak – it is likely that this is one more ability of the two-dimensional wonder material that may come in useful in the future alongside its other properties. It's possible to imagine that a graphene film might be floated over an air gap, for example, using the variable magnetic field of an electromagnet to fine tune its interaction with another material. This would make it possible to vary the charge held in a capacitor at the touch of a button, making it easier to control the speed at which a supercapacitor replacement for a battery (see page 136) discharges.

Entertaining though the idea of floating strips of graphene may be, there are likely to be much more significant applications that derive not so much from its electrical properties, or even its strength, but as a result of graphene and the other ultrathin material providing such a unique atomic lattice.

Thin drinks

Specifically, the special nature of two-dimensional materials can provide benefits as a result of the way the two-dimensional lattices interact with other atoms and molecules.

This becomes apparent in developments being made to use these materials in the desalination of water. If it can be done simply and cheaply, desalination is a hugely important technology. The Earth is not exactly short of water. There's about 1.4 billion cubic kilometres of the stuff. A single cubic kilometre contains a trillion litres of water. Yet areas of the world can be chronically short of drinking water, even if they have ready access to the sea – because the vast majority of those cubic kilometres are undrinkable salt water.

Desalination has been possible for a long time, either by variants on distillation or by using special membranes where pressure is used to force water through a material that holds the salts back in a process known as reverse osmosis, as mentioned above (page 108). However, the reverse osmosis method takes a considerable amount of power to run and features expensive membranes that have to be regularly cleaned or replaced. Molecular sieves based on ultrathin materials could be a far cheaper option. A first approach was to use graphene oxide membranes parallel to each other, so the gap between was small enough to let water through, but not big enough to allow the salts in the water to pass through.

This approach worked for larger salts and contaminants, but small salts – though they shouldn't have been able to get through the gaps – stayed in the water. It was discovered that when the membranes were soaked in water they swelled sufficiently to let the smaller salts through some of the channels. With careful adjustments of the size, it has proved possible to make a largely effective molecular sieve for desalination. However, this wasn't the end of the development.

A more controlled approach would be to make just the right sized holes or slits in a material to let only the water through – but this doesn't usually work with conventional membrane materials, because the surfaces of the membrane are too uneven to be able to keep the incisions correctly sized. However, high-quality two-dimensional materials don't have such irregularities – there is no variation in the surface.

Researchers at the Graphene Institute in Manchester have managed to produce slits less than a nanometre across using graphene, boron nitride and molybdenum disulfide sheets. These gaps are of the same scale as the molecules such as water that are being worked with for filtration. Cutting slits so accurately would prove technically challenging to say the least, but Andre Geim and his team came up with a clever alternative.

They produced two thin pieces of graphite, each with atomically smooth surfaces. These crystals, around 100 nanometres thick, would form either side of the slit. They then placed strips of two-dimensional material along two parallel edges of one of the crystals and rested the other crystal on top, producing a sandwich. The result was a pair of crystals, wedged apart by a gap no thicker than the two-dimensional material used. As Geim explained:

> 'It's like taking a book, placing two matchsticks on each of its edges and then putting another book on top. This creates a gap between the books' surfaces with the height of the gap being equal to the matches' thickness. In our case, the books are the atomically flat graphite crystals and the matchsticks the graphene or molybdenum disulfide monolayers.'

The whole structure is held together by van der Waals forces and the size of the slits is similar to the tiny gaps provided in living cells by proteins called aquaporins, which allow water and ions* to move through cell walls, a process that is essential for biological functions. When an electrical potential difference is applied from one side of the molecular

---

* As we have seen, ions are atoms which have gained or lost one or more electrons and become electrically charged. For example, in seawater, salt is not present as sodium chloride, but as positively charged sodium ions and negatively charged chlorine ions.

sieve to the other, different ions move through the slits. Surprisingly, ions bigger than the slits can get through them, as atoms aren't solid balls but have a degree of flexibility. The hope is that with a better understanding of how such slits can be used to control ion movement it will be possible to produce desalination plants that can process seawater much more quickly and with far less power consumption than would be possible using a simple membrane sieve approach.

In this case, the sieve action came from the slits between the blocks of graphite, held apart by ultrathin strips. However, a variant of graphene has provided an alternative approach that produced the remarkable property of automatically distilling liquor without any added energy required.

The magic still

Graphene continues to surprise those working on it with this kind of unexpected new trick. In 2012, Andre Geim's team reported one of its strangest behaviours yet. Because of its continuous lattice structure, graphene is very good at keeping liquids and gases at bay. The team produced sheets of multi-layered graphene oxide – graphene with 'hydroxyl' structures (OH) randomly attached to the surface, forming a self-supporting membrane, still hundreds of times thinner than a human hair.

When the membrane was used to line a metal container it proved superbly effective at preventing liquids and gases escaping – even helium, which is notoriously hard to keep in place. This is impressive when you consider that a millimetre-thick coating of glass won't stop helium slowly working its way through – but this wafer-thin membrane kept it in place.

Of itself, this wasn't too much of a surprise, as graphene's structure doesn't allow much in the way of escape routes. But what was remarkable was that one thing could get through the graphene oxide membrane. Water. Liquid water couldn't be poured through the membrane, but the team discovered that water would evaporate through it at the same rate as it evaporates to open air. This was a stunning discovery.

What appears to be happening is that graphene oxide sheets have just the right spacing between them for a single molecule-thick layer of water to slide through the gap. Bigger atoms or molecules don't fit and smaller ones, such as helium, cause the structure to shrink and close up the space – only water seems to have the magic touch.

This being Andre Geim's team, there was an irresistible application that suggested itself. If you place a water-based mixture into a container sealed with this membrane, over time the water will evaporate off, leaving the rest of the contents behind. One of Geim's team, Rahul Nair, commented: 'Just for a laugh, we sealed a bottle of vodka with our membranes and found that the distilled solution became stronger and stronger with time.' Nair himself doesn't drink, but it's hard to imagine that the super-strong vodka went to waste.

While self-distilling alcoholic drinks have a certain appeal, there are plenty of other applications where it may be useful to reduce the water content of a mixture without allowing other volatile substances to escape, for example to remove contaminants from fuels, or to reduce water content of fruit juices for transport without losing any of the volatile compounds that give them the 'freshly squeezed' taste.

Moving away from liquids, though, one of the original graphene team has found another use for graphene that may be niche, but plays to its strengths.

The microscope's friend

One of the smaller potential applications of graphene, but one that Konstantin Novoselov is fond of, is as a support structure for materials to be examined when using transmission electron microscopes – the kind of electron microscope wherein a stream of electrons – fulfilling the role performed by light in a traditional optical microscope – is passed through a material rather than reflecting off it as happens with a scanning electron microscope. What's needed to support the sample is a strong material that isn't damaged by radiation and that is a good conductor of electrons – and graphene ticks all the boxes.

The process would involve transferring a layer of graphene onto one of the metal support grids used in transmission electron microscopy. It could be exposed to the substance to be studied in solution and would support and hold in place the biological material or other substance far more consistently than a traditional electron microscope slide.

This kind of application is likely to have some indirect medical use, but there is a way that graphene could have a much more direct benefit for medical teams, by providing diagnostic tattoos.

What use is an invisible tattoo?

There is considerable excitement in medical circles over the possibility of creating graphene tattoos. These are not the latest fashion in body modification – after all, graphene is pretty much transparent, so graphene tattoos are hardly showy – but rather a remarkable potential mechanism for producing health-monitoring devices that will hardly be noticed by the user, even when engaging in movements that would tend to rip away an ordinary sensor.

Although the term 'tattoo' sounds worryingly permanent, the graphene tattoos are not inserted under the skin, but use the same adhesive technology as temporary tattoos. Lasting around two days before they fall away, they can easily be removed earlier if required, by the familiar graphene transport trick of applying a piece of sticky tape.

In comparison to the kind of sensor found in fitness bands or smart watches, a graphene tattoo has the huge advantage of contorting to match the shape of the skin as the wearer moves about. This means that the electrical contact stays constantly in place, a requirement for medical-grade data, whether the tattoo is being used to monitor heart rate or bioimpedance (the skin's electrical resistance), the latter being used both to give an idea of body composition levels – a more effective measure of body fat levels than BMI – and diagnostically for a range of cardiac, pulmonary, renal, neural and infection-based disorders.

The constant contact, thinness and flexibility mean that a graphene tattoo would be significantly less obtrusive than a fitness band but would have at least as good a connection as existing medical sensors, which have to be glued on with an application of conducting gel. The traditional sensors can cause skin damage when removed, particularly to elderly patients, and are relatively expensive to produce.

As we've seen many times already, graphene needs some kind of substrate to prevent it from crinkling up. In the graphene tattoo, that substrate sits above the graphene, rather than below it, in the form of a transparent polymer called polymethyl methacrylate (PMMA for short). The structure of the sensor is then laser-cut onto temporary tattoo paper, which is used to apply it to the skin.

Not only is the graphene tattoo nearly transparent, it is as flexible as human skin, so is hardly felt by the wearer, even when covering relatively extensive areas. In the first practical tests of the technology in 2017, the prototype tattoos were used to check skin temperature and hydration (measured from skin conductivity) and as electrodes for electrocardiograms, electromyograms and electroencephalograms, measuring activity in the heart, muscles and brain.

This flexibility makes graphene technology a natural for something attached to the skin, but there are plenty of other circumstances where having non-rigid electronics could be useful – and potentially doing far more than just acting as an invisible electrode.

Flexible fun fashion

The flexibility of graphene and the other thin film materials gives them a natural potential for producing wearable electronics. This is the same kind of technology as the graphene tattoos discussed in the previous section, but in this case could be applied either to the skin or to clothing, as a membrane like the tattoo or as graphene ink for less serious applications. It's possible that such wearable electronics could enable anything from direct interaction between human and machine – controlling a machine by gestures, for example – to using the light-emitting capabilities of molybdenum disulfide to produce clothing that lights up with a still or moving image.

The researchers who were responsible for the graphene tattoo suggest that wearable graphene electronics could be used for interaction with smart houses and controlling wheelchairs and robots. They have already demonstrated the ability to control a drone wirelessly using signals from a graphene tattoo.

Another 2017 development was at the Tsinghua University in Beijing, where a graphene-based strain sensor was used to change the colouring of a layer above it. A layer of graphene acted both as a very sensitive measure of strain – since distorting the graphene lattice changes its electronic properties – and as an electrode to control an organic electrochromic device. This is a material that changes colour depending on the electrical voltage applied to it.

The result was a thin, flexible sheet of material giving a constant visual readout on the amount of strain applied to the sheet. The visual strain sensor has both potentially entertaining applications – for example, clothing that changes colour with your movements – and practical use as a sensitive strain gauge that could be used to monitor conditions in anything from high-risk construction to medical swelling.

As far as electronically enhanced clothing goes, in 2017, researchers at Cambridge, working with colleagues from Italy and China, produced fabric which had graphene circuitry directly printed onto the fibres of the material. The result was electronic circuitry which is as flexible as the garment and can survive up to twenty washes.

This development reflects one of the most significant steps in these kinds of graphene technology – the development of inks based on graphene and some of the other two-dimensional substances, which can be printed onto a fabric (or a piece of paper) from an inkjet printer. Even at this very early stage, the researchers have been able to produce fully functional, all-printed electronic circuits. Although inks for printed electronics had already been used to a limited extent, they use solvents that are dangerous for human contact, and are not effective on flexible materials. However, the materials based on layers of graphene and boron nitride have the usual flexibility of two-dimensional materials and are non-toxic and environmentally friendly.

Most importantly, the visual side is not all that can be provided using printable two-dimensional inks. Anyone who has ever worn a shirt with built-in technology for lights or other devices knows that the real problem is not so much the display as the power source.* They come with a separate, clunky battery pack which has to be housed somewhere when the garment is worn, and removed for washing. However, in

---

* I was once given a shirt that lit up with a Wi-Fi power level symbol when it was in the presence of a Wi-Fi hub, displaying how good the signal was. The concept was great, but for the reasons mentioned here it was a pain to wear. It was retired after one go.

2017, the Graphene Institute demonstrated supercapacitors (the ultrathin alternative to batteries – see the next section) produced using graphene oxide ink printed on to cotton fabric. The cotton fibres act as a substrate for the graphene oxide supercapacitors, which remain as flexible as the fabric. As well as making clothing-based electronics more practical, this would be equally valuable used in conjunction with the graphene tattoos to power medical diagnostic and fitness monitoring devices.

Although making a T-shirt light up is an entertaining use of ultrathin materials, the supercapacitor itself has the potential to do far more. It could even be the answer to the electric car problem.

Supercharged stores

One of the biggest drawbacks of our increasing dependence on batteries – whether it's to power our smartphones or, at the other extreme, an electric car – is the time that it takes a battery to charge. Thanks to graphene's electronic abilities, a team at the University of Waterloo have managed to make significant steps forward in the production of supercapacitors, which are devices that have the potential to replace batteries, but can be charged up in seconds.

A capacitor (the electronic component formerly known as a condenser) is a device that holds electrical energy in the form of an electrical field, as opposed to the chemical energy used for storage in a battery. This means that a capacitor can charge up and discharge extremely quickly – as quickly as you can get the power into it, without having to wait for a gradual chemical process to take place. At its simplest, a capacitor is just a pair of metal plates with a sheet of dielectric medium between them, usually a type of plastic. 'Dielectric' means that it is an insulator, but is able to hold an electrical charge – so although it doesn't allow a current to flow between the plates, the dielectric becomes positively charged on one side and negatively charged on the other.

Once a capacitor is charged up, it can hold that charge for some time before it is connected to a circuit and the charge is released. This ability makes capacitors a common component in electrical circuits, particularly as they will block a direct current from flowing, but will allow an alternating current through. Direct current (DC) is the kind of continuous electrical current in one direction produced by a battery, while alternating current (AC) is the type of electricity used in mains supplies, where the direction of flow alternates, its voltage usually varying like a sine wave. Signals in electrical form usually come in the form of waves and one of the benefits of the capacitor's ability to lose DC is to filter out unwanted noise to leave the pure wave behind. But the capacitors face two problems if they are to replace batteries – the amount of charge they can hold and the speed with which they discharge it.

Just as a capacitor can be charged up far quicker than a battery, it can also let all its charge go as quickly as the resistance in the circuit allows it. When I constructed a laser flashtube in a sixth form project at school, we used capacitors borrowed from Manchester University which were the size of five-litre oil cans, could hold enough charge to kill, and discharged in a fraction of a second to make a flash of miniature lightning several centimetres long.

However, the industry has plenty of experience at making circuitry that will gradually release a capacitor's charge – it just takes more than the capacitor alone. The other problem, capacity, has proved more intractable. Until recently, the amount of charge a capacitor can hold has not been sufficient to make them a useful replacement for a battery.

Supercapacitors, which (as the name suggests) store far more energy than conventional capacitors, are already in use, for example in electric cars using regenerative braking where the electrical energy produced during braking has to be rapidly stored. However, they are relatively bulky and limited by the ability of their electrodes to handle electrical currents. Graphene has the potential to facilitate improvement on existing designs.

At the moment, a supercapacitor that would fit into a mobile phone could only hold around 10 per cent of the charge of an equivalent-sized battery. But by giving the supercapacitor ultra-high-capacity electrodes, its storage can be significantly increased. The Waterloo team have made their prototype supercapacitors from multiple layers of graphene sheets, separated by an oily liquid salt. This medium prevents the graphene sheets from directly interacting and losing their two-dimensional properties – resulting in a multi-layer electrode that can handle far more electrons. The role is not just structural, though – the oily liquid salt provides the dielectric medium for the supercapacitor – so the whole component becomes far smaller in size and weight than with conventional materials.

Although with current designs it's unlikely that super-capacitors will entirely replace batteries, it's certainly a possibility for the future, as we are still very much at the start of developing ultrathin technology. Think how long it has taken us to get to our current battery capability – yet graphene has only been in practical use a handful of years. Within a relatively short timescale, graphene-based super-capacitors could give us phones that charge in seconds, and the ability to top up an electric vehicle's power in less time than it takes to fill a tank with petrol. If that were to happen, what is arguably the biggest drawback of electric cars will be removed and we can look forward to far greener transport.

At the other extreme of scale to supercapacitors, tiny graphene capacitors have already been used as a way of detecting previously undetectably small changes in pressure.

Under pressure

In an ultrathin pressure sensor, graphene sheets on a thin polymer substrate are arranged across shallow depressions in a silicon chip. Any change in pressure on the graphene membrane changes the distance between the graphene and the silicon, closing the gap. In combination, the graphene and the air gap act as a capacitor: as the distance between the two substances changes minutely, the capacitance of the device changes too, which can be measured in a circuit.

As well as providing an alternative approach to the Sussex touch screen mentioned above, this type of sensor could be used in future to monitor engines, heating, ventilation, air conditioning and more, with remarkable sensitivity.

Before we leave the topic of energy, graphene has one more surprise to reveal. It isn't just able to store energy, it can generate it too.

The smallest generator

We have already seen (page 86) that before Geim and Novoselov's work, it had been expected that graphene would be unstable, ripping itself into blobs of carbon under the influence of the natural movements of its atoms. However, a combination of its unexpectedly large tensile strength and the restraining van der Waals forces when it is based on a substrate have tended to keep it in check. But a team at the University of Arkansas have not only spotted some residual action in the graphene atoms; they also believe that they can harness the energy of movement to produce a tiny electrical generator.

It all started when they were observing graphene under an electron microscope on a copper substrate. The image they got initially was confusingly unclear. It was only when they managed to break down what was happening in time and space that they realised they were seeing the carbon atoms fluctuating up and down in a random rippling. This was made up of a particular type of random distribution that has occasional much larger jumps, known as a Lévy flight. The rippling was the result of the heat in the room providing energy to keep the carbon atoms in energetic motion.

All the molecules in a liquid or gas are also in random motion, of course, moving far faster than the atoms in a solid. We see this in Brownian motion – the way that tiny particles of matter, if suspended in water, are bounced around by the constant barrage of moving water molecules. But the big difference is that the usual random motion can't be harnessed because all the different motions in different directions cancel each other out. The motion of those particles is a truly random walk. Things are different, though, with graphene.

Because all the atoms are in a single lattice in a sheet, they are constrained to move more together, and rather than producing a self-cancelling collection of tiny movements, the movements occur in similar directions causing relatively large-scale ripples in the graphene, which is where it is hoped energy can be produced.

There is still some concern here that what works in theory might not work in practice. While a prototype nano-generator has been designed, it has yet to be constructed and tested. However, in principle this could mean that devices requiring small amounts of power could be kept running indefinitely by a tiny power source that never needs replacing.

Note, by the way, that this not yet another perpetual motion fantasy or a variant on Maxwell's demon.* The energy

---

* The demon is a thought experiment which supposedly defies the second law of thermodynamics by allowing heat to pass from a cooler to a warmer place, using a tiny demon that can see which air molecules are fast and which are slow, and using a partition to separate them into different containers. Unlike the fictional Maxwell's demon, the graphene generator does not break the second law of thermodynamics.

is not being produced from nowhere. Using the device would slightly cool its surroundings as it is the thermal energy of the graphene's surroundings, ultimately originating from sunlight, that drives it. It is, in effect, an indirect solar engine making use of ambient heat.

The only thing that makes this unlikely behaviour possible is the remarkable strength of graphene, which prevents the thermal fluctuations from tearing it apart. But clearly there are other benefits to be gained from the super strength.

Super strength in action

As we've already seen, graphene is by far the strongest substance as far as tensile strength goes. This means that it is almost certain to have a role in applications where strength is key, and materials are being developed with multiple strips of graphene embedded in different kinds of polymer. Wherever, for example, carbon fibres have been used to date, if even they aren't strong enough, graphene is likely to take over.

Of course, some strength-related applications are liable to be more spin than substance. Being able to say that you have the strongest material ever discovered in your product could easily give it an apparent edge that thin slivers of graphene may not realistically deliver. For example, running shoe manufacturer Inov-8 has designed a new 'G-series' running shoe in collaboration with the Graphene Institute. Due out in 2018, the G-series will have graphene included in their rubber soles. The sportswear manufacturer claims that it will 'make the shoes stronger, stretchier and more durable'. It might seem this is a running shoe equivalent of cosmetic magic ingredients such as Pro-Retinol A, and we are certainly likely to see graphene turning up in quite a few products where its presence is more about the name than anything else. However, the contribution of the Institute team does make the running shoe claims seem more likely, and indeed Manchester researcher Aravind Vijayaraghavan claims that 'The graphene-enhanced rubber can flex and grip to all surfaces more effectively, without wearing down quickly, providing reliably strong, long-lasting grip.'

Even though a certain amount of flummery may well be in store, though, it's worth remembering that graphene is not a brand name or just a bit of 'science stuff' to make trainers seem more impressive. It is, without doubt, a true wonder material.

Pinch me

The previous section on graphene's strength was going to be my last example of graphene applications, but this is a topic where everywhere you turn something new is being tried – and one of the latest announcements at the time of writing was yet another totally different way to use graphene. Researchers working at the University of Minnesota's College of Science and Engineering have found a way to produce graphene tweezers so small that they can grab hold of individual biological molecules floating in water.

What's happening here has been done for some while, using a process called dielectrophoresis – but it was incapable of working at the scale of single molecules. The mechanism works by holding the item to be trapped in an intense electrical field. Because graphene is so thin, coupled with its superb conductivity, using its edge to produce this pinpoint localised field is far more precise that has otherwise been possible.

Because the graphene electrodes are so narrow, the voltage required to set up the field is small. In conventional dielectrophoresis, high voltages that are only usable under controlled laboratory conditions are required. But the graphene electrodes could grab, for example, a molecule of DNA with only 1 volt needed. Not only does this make the procedure much safer, it also means in principle that it's possible to imagine medical diagnostics devices linked to a smartphone that could use this technique to isolate biological molecules and analyse them for diagnosis of diseases.

The analysis part does not even require a separate device. The tweezers can not only be used to grab a molecule but to act as extremely sensitive biological sensors which could enable the same tiny piece of kit to provide the data for a range of diagnostics. The Graphene Express goes on and on.

And there's more

All of the above might seem more than enough – far more than anyone could have imagined when graphene was first discovered. From our experience with carbon fibres, it's possible that graphene's tensile strength could have been predicted. And there was already a suspicion that its electronic capabilities might have been interesting. But no one could have guessed just how massive a change graphene and the other two-dimensional substances would bring. And those discoveries keep coming. Look at one of the online resources at the back of the book and you will likely find a whole string of new graphene discoveries – often all made in recent months. This is not a single breakthrough but a massive chain of events.

In a sense, graphene has ushered in a whole new type of materials science dealing with these ultrathin materials. When researching this book, I could have filled a whole chapter with just the innovations discovered in one year. With their flexibility, transparency, strength and small scale, ultrathin materials are appealing to thousands of researchers across the world. It's almost as if we had been given a whole new set of elements – certainly of structural building blocks – and were able to start manufacturing devices that were previously inconceivable.

All this started with those Friday night experiments in Manchester. And just as the capabilities of ultrathin materials have blossomed, so have the facilities and staff involved in this work.

From backroom to mainstream

Beginning with those initial small, spare-time experiments in Manchester, work on graphene and other ultrathin materials has seen a worldwide boom, with activities under way in all the major research countries, recognising the technological promise of these materials. A decade on from the publication of Geim and Novoselov's discovery in 2004 – a very short time in terms of such a major technological breakthrough – activity had become intense, and it only continues to grow. To see how much has been made possible by graphene, we only have to look again at developments at Manchester University.

In Manchester, Geim and Novoselov originally worked in a small clean room in a corner of an old building on the campus. As interest grew, the university found that the expansion of work on graphene and other two-dimensional structures was taking off at a far faster rate than their available capacity could cope with. Within a few years of the pair winning their Nobel Prizes in 2010, there were 30 professors and around 200 students working in Manchester, all dedicated to this field. With accommodation in existing buildings stretched at best, the university got grants to build a new National Graphene Institute on a spare plot of land on the Manchester campus.

The specification was, to say the least, challenging. Konstantin Novoselov has described the Institute as 'a building of probably unrivalled complexity'. Not only did it require large numbers of clean rooms for working in a totally dust- and contamination-free environment,* it was being built for equipment that was as yet not even designed and for purposes that would not become clear until research

---

* The contamination-free requirement is a pleasing parallel with the original physics laboratory at Manchester that Rutherford had worked in, with its air purifiers, drawing the smoky industrial city air over oil baths to clean it.

moved further on. One thing that is certain with graphene research is that it will always feature surprises.

Construction began on the Institute in 2013, based on a complex architectural specification. Not only would the clean rooms have to be protected from air pollution, it was also essential that vibration was kept to a minimum, as sensitive devices such as scanning tunnelling microscopes, and the whole business of manipulating materials at the nanoscale, were easy to disrupt. The Institute building was to be located on a main road, a constant source of vibration, so the main clean room was pushed five metres below ground level, enabling it to be anchored directly to the bedrock below. Similarly, the section of the Institute containing heating and air conditioning units, which inevitably cause some vibration themselves, was given a clear 50-millimetre insulating gap to separate it from the rest of the building.

In a way, Geim and Novoselov's Friday night thinking approach proved both a bonus and a challenge for the builders of the Institute. Novoselov, who was the more involved of the two in the design, made it clear that the optical, electronic, chemical and general laboratories needed to be easily adaptable for future unknown experiments and wanted offices that were interspersed with the labs, so there was easier sharing of information and ability to take in different viewpoints, rather than taking the architecturally simpler approach of putting all the offices together, separated from the lab work. The building was also designed to have a biodiverse roof garden with grasses and wildflowers to attract bees and other pollinators.

The National Graphene Institute building was fully opened in 2015 and is to be followed by two further developments. At the time of writing, construction is nearly complete on the institute's sister Graphene Engineering Innovation Centre, which is due to open in mid-2018. This is designed to provide the next step for the original research that emerges from the Institute, putting many of the concepts mentioned in the previous pages into prototyping for production, ironing out the engineering challenges that always arise when going from the lab to the factory.

The Innovation Centre will be followed a year later by the significantly larger-scale Sir Henry Royce Institute for Advanced Materials Research, providing a centre for work on ultrathin materials as well as a range of other specialist material science developments. With its hub sited across the road from the Graphene Institute, the Royce Institute is a joint partnership between the universities of Manchester, Sheffield, Leeds, Liverpool, Cambridge, Oxford and Imperial College London, cementing Manchester's position as the world's leading location for ultrathin development – all because of Geim and Novoselov's original way of thinking.

Patents and Flagships

Although the work in Manchester is world-leading, the scope and potential benefits of graphene and the other ultrathin materials are so wide-reaching that large sums are being invested across the world to advance research and bring products to market. This is reflected in the graphene-based patents issued. In 2007, 161 patents were issued. By 2009/10 the rate was up to around 1,000 a year. This rise continued to a peak of nearly 7,000 patents in 2015 – this dropped in 2016, but at the time of writing it's too early to know if this was just a short-term deviation.

The business of patenting here is a complex one. Graphene itself couldn't be patented – it's a natural phenomenon – and European companies and universities and businesses have been slow to take out patents. By far the most prolific collectors of patents are China and Korea, with the US next. Many of the patents will never result in practical outputs, and some are regarded as cynical business ploys in a process known as patent trolling, where companies issue as widespread patents as they can in the hope of catching some future development in their net.

It's also the case that around 98 per cent of the Chinese patents, for example, only cover China and seem to be due to a quota system that drives a large number of dead-end patent applications. What is interesting is that when the number of graphene-producing organisations is compared, while China still comes top, the US is next, followed by the UK. More interesting still, we have to bear in mind how the sheer size of the Chinese and US economies distorts absolute figures. When the graphene producers are taken as a proportion of GDP, Spain tops the table with the UK in second place, then India, pushing China into fourth. Countries such as Spain and the UK seem to be performing better than might be expected because there has been a conscious funding bias towards graphene developments.

There is no doubt that there will be significant product development from the patent-hungry countries – but as we have seen, a big investment is being made in research in the UK, and this is matched by a wider push in Europe through a collaboration known as the Graphene Flagship. This pulls together input from over 150 academic and industrial research groups in 23 mostly – though not uniquely – European Union countries. Although the Graphene Flagship is inevitably more bureaucratic in operation than a single institute, it has the ability to pull together information on an unprecedented scale.

One example of its work is the testing done in 2017 in collaboration with the European Space Agency to explore opportunities for graphene in 'space-like' applications. One experiment looked at making use of graphene's excellent conductivity in a 'loop heat pipe' – effectively a cooling system that transfers heat from the hot pipe into a liquid. Although graphene's abilities are well recorded in normal conditions, for space applications there is the need to check out if there is any change either under zero gravity or the heavy acceleration of lift-off. To date, all the evidence is that graphene will continue to perform just as effectively.

Another experiment made use of graphene's combination of strength and thinness in testing out solar sails – space-borne sails which pick up the pressure of the light from the Sun to accelerate a craft – made of graphene membranes. Again, this had to be performed in effectively zero gravity – achieved using the 146-metre drop tower in Bremen, Germany, which allows an experiment to undergo 9.3 seconds of weightless free fall as it plummets down the tower. More work is needed here, but again the indications are good.

Creativity in action

All in all, Geim and Novoselov's work on graphene is a wonderful example of creativity at work in a scientific sphere. Geim is convinced that his approach of encouraging Friday night experiments provides a mechanism to ensure that scientists do not get stuck in a rut, as can easily happen with their usual extremely narrow focus. There are interesting parallels with the approach taken for decades at the US company 3M, in which engineers are encouraged to spend half a day a week working on something that isn't their main activity – a fun side project that could result in a whole new product for the company. With a pleasing similarity to the idea of peeling off layers of carbon in graphene, Post-it Notes are just one of the products to emerge from that 3M process.

All too often, scientists spend all their time focused on a single detail in a very narrow field. But the Friday night experiment approach really does seem to provide an opportunity that many more scientists could benefit from. Geim notes that when preparing for his Nobel lecture, he compiled a list of around two dozen Friday night experiments he and his colleagues had undertaken over the years. Most, as might be expected, had failed. Failure is an important part of the creative process. But there were three hits, in levitation, Gecko tape and graphene. As Geim points out, this makes for an impressive success rate of better than 10 per cent, bearing in mind these were just fun little projects with very little budget.

There were also a number of near misses among the apparent failures, showing how remarkably productive the process is. Geim believes that this success rate is not because the ideas (or even the scientists) were particularly brilliant, but because 'poking around in new directions, even randomly, is more rewarding than is generally perceived'. You may well fail in this kind of venture, but by limiting it to a relatively small part of work time and allowing the imagination free rein, it seems a very powerful way to ensure that new paths are explored, sometimes leading to a remarkable reward, as happened with graphene. And, as Geim says, one is at least guaranteed an adventure.

A second plank of Geim's approach, which might surprise some of his peers, is not to put too much effort into checking the literature for other people's attempts in related fields. Yes, he suggests, it's still important to have a couple of reviews to ensure that your exciting new idea is not entirely reinventing the wheel, but if you spend all your time looking at the literature and not trying things, Geim claims that you won't do anything useful – and will no doubt come to the conclusion that your idea has been tried before and it didn't work, so there's no point going any further. But every attempt is subtly different and sometimes it only takes a small factor to make a big difference.

Bear in mind that it was well established before Geim and Novoselov's work on graphene that two-dimensional structures of this kind *could not be made*. They were 'known' to be unstable. But the accepted 'facts' were wrong. Nothing is going to be entirely new. But taking a different approach, looking at things a different way, can make all the difference. And even with that determination to give it a try, without the inspiration of the dirty tape in the waste bins, that happy coincidence, nothing might have come of that particular Friday night project.

Geim and Novoselov could have taken note of the theory that said it wasn't possible to make stable graphene. But they didn't.

And the outcome is quietly changing the world.

FURTHER READING

There has been relatively little published for the public on graphene, other ultrathin materials, and their applications. This means that often the best opportunities for further reading will come from websites and other more transient media rather than from books.

Chapter 1: The sticky tape solution

The Graphene Institute website: www.graphene.manchester.
ac.uk – elegant website with some information on the
people involved, the work at Manchester, graphene itself and
applications news.

Andre Geim's Nobel lecture: www.nobelprize.org/nobel_prizes/
physics/laureates/2010/geim_lecture.pdf – a very readable
piece by Geim on his approach to science, his personal
history and the discovery of graphene.

Konstantin Novoselov's Nobel lecture: www.nobelprize.org/
nobel_prizes/physics/laureates/2010/novoselov_lecture.pdf
– a more technical piece than Geim's, but largely readable on
the nature of graphene and some of its potential applications.

Chapter 2: The essence of matter
Atom, Piers Bizony (Icon, 2017) – a good introduction to our
    gradual understanding of the nature of matter.

Chapter 3: Quantum reality
The Quantum Age, Brian Clegg (Icon, 2015) – a guide to quantum
    physics with more information than is generally provided on
    applications, from lasers to electronics.
Cracking Quantum Physics, Brian Clegg (Cassell, 2017) – a highly
    illustrated introduction to the basics of quantum physics.

Chapter 4: Like nothing we've seen before/
Chapter 5: Other flatties
Graphene: A New Paradigm in Condensed Matter and Device Physics,
    E.L. Wolf (Oxford University Press, 2014) – not much use
    unless you have a physics degree, but if you can take the
    technical content this provides a good way to discover why
    graphene and the other ultrathin materials are so special.
National Graphene Institute News: www.graphene.manchester
    .ac.uk/latest – a good way to pick up on some of the latest
    developments in two-dimensional material applications,
    though inevitably biased to those discovered in Manchester.
Science Daily Graphene News: www.sciencedaily.com/news/
    matter_energy/graphene – the latest happenings in graphene
    from around the world.

Chapter 6: The ultrathin world
The Right Formula: The Story of the National Graphene Institute, David
    Taylor (Manchester University Press, 2016) – a bit of a glossy
    sales brochure, but still has interesting snippets on both the
    discovery of graphene and the construction of the Institute.
Graphene Flagship – more information about the Graphene
    Flagship consortium and other breakthroughs in graphene
    can be found on the website graphene-flagship.eu

Graphene patents – an overview of the state of patents in 2015 can be found in the UK Intellectual Property Office report 'Graphene: the worldwide patent landscape in 2015', available at www.gov.uk/government/uploads/system/uploads/attachment_data/file/470918/Graphene_-_the_worldwide_patent_landscape_in_2015.pdf

INDEX

Asterisks indicate footnotes

3M 151

A

Abrahams, Marc 4**
Agrigentum 19
Anderson, Carl
    discovery of positron 62
anti-electron *see* positron
Aristotle 19–21
atoms
    allotropes of carbon 38–9
    bonding 34–6, 47
    hexagonal form of ice 37
    hydrogen bonding 37
    shells around atom 33, 34, 55

B

Balmer, Jakob 54
    Balmer's equation 55
band gaps
    bigraphene 117
    'conduction band' 56–7, 64, 92
    semiconductor band gap 56, 63
    silicene 109
    'valence band' 56, 57, 64, 92
    'zero band gap' 57

Bardeen, John 76
Bernal arrangement 117
Berry, Michael
    collaboration with Geim on frog
        project 5, 8
Berzelius, Jöns Jacob 25
Big Bang Theory, The 93*
Bohr, Niels 3
    Carlsberg Foundation grant 52
    exploring structure of atom 52–5
    letter to brother Harald 53
    quantum model of the atom
        32–3
    Rutherford, working with 52
    study in England 51–2
    Thomson, working with 52
Boole, George
    Boolean algebra 66
Born, Max 47–9
boron nitride
    applications 105–6
    hexagonal lattice 104
    shared properties with graphene
        105
    substrates, use in 80
    visual electronics, use in 135